Sign Language and Language Acquisition in Man and Ape

New Dimensions in
Comparative Pedolinguistics

AAAS Selected Symposia Series

Published by Westview Press
5500 Central Avenue, Boulder, Colorado

for the

American Association for the Advancement of Science
1776 Massachusetts Ave., N.W., Washington, D.C.

Sign Language and Language Acquisition in Man and Ape

New Dimensions in Comparative Pedolinguistics

Edited by Fred C. C. Peng

AAAS Selected Symposium **16**

AAAS Selected Symposia Series

Copyright © 1978 by the American Association for the
Advancement of Science

Published in 1978 in the United States of America by
   Westview Press, Inc.
   5500 Central Avenue
   Boulder, Colorado 80301
   Frederick A. Praeger, Publisher

Library of Congress Catalog Card Number: 78-19807
ISBN: 0-89158-445-5

Printed and bound in the United States of America

About the Book

This volume brings together recent research findings on
sign language and primatology and offers a novel approach to
comparative language acquisition. The contributors are anthropologists, psychologists, linguists, psycholinguists, and
manual language experts. They present a lucid account of what
sign language is in relation to oral language, and of the capacity of nonhuman primates to learn specific, human-origin
language systems. The papers give a contrastive view of man
and ape, as well as a general overview of comparative language
acquisition.

In the first half of the volume, sign language is put in a
cultural context and assessed from historical and contemporary
perspectives. It is seen not as a self-contained entity but
rather as a general norm of individual sign activities that
rapidly develop and change in any given society. Moreover,
sign language, a system of the visual mode and the brachial
apparatus, is shown to be useful for measuring the similarities
and differences between man and ape.

The second half of the volume is concerned with the assessment of research in which signing apes are compared with
humans and with each other, using sign language (or some other
visual sign mode) as a yardstick. The purpose of this comparison is the search for a common precursor of human and nonhuman communication systems, and ultimately, the discovery of
a general faculty shared by man and ape.

Contents

List of Figures

List of Tables

Foreword

The *AAAS Selected Symposia Series* was begun in 1977 to
provide a means for more permanently recording and more
widely disseminating some of the valuable material which is
discussed at the AAAS Annual National Meetings. The volumes
in this *Series* are based on symposia held at the Meetings
which address topics of current and continuing significance,
both within and among the sciences, and in the areas in which
science and technology impact on public policy. The *Series*
format is designed to provide for rapid dissemination of information, so the papers are not typeset but are reproduced
directly from the camera copy submitted by the authors, without copy editing. The papers are reviewed and edited by
the symposia organizers who then become the editors of the
various volumes. Most papers published in this *Series* are
original contributions which have not been previously published, although in some cases additional papers from other
sources have been added by an editor to provide a more comprehensive view of a particular topic. Symposia may be reports of new research or reviews of established work, particularly work of an interdisciplinary nature, since the AAAS
Annual Meeting typically embraces the full range of the
sciences and their societal implications.

WILLIAM D. CAREY
Executive Officer
American Association for
the Advancement of Science

About the Editor and Authors

Fred C. C. Peng, professor of linguistics at the International Christian University in Tokyo, has studied many areas of linguistics and has published seven books and more than sixty articles, including On the Nature of Sign Language *and* Aspects of Sign Language *(in Japanese; Bunka Hyoron Publishing Co., 1976 and 1977). He has lectured throughout the United States and Japan and his positions include secretary of the International Association for the Study of Child Language, director of the International Christian University Language Sciences Summer Institute, and secretary and co-founder of the Sign Language Society of Japan. Dr. Peng also holds editorial positions with five professional journals.*

Richard Blasdell, a research associate at the Manual Language Department of the National Technical Institute for the Deaf, is currently conducting a program of research into the physiological aspects of manual communications and speech in normal-hearing and deaf adults. A specialist in psycholinguistics and language acquisition, he is the author of numerous articles on manual language and other subjects.

Charles Bradley, a graduate student in physics at Washington University, was previously an interpreter at the National Technical Institute for the Deaf. He is a member of the Registry of Interpreters for the Deaf, with comprehensive certification, and he has published an article on technical signs standardization in the American Annals of the Deaf *(February, 1977).*

Frank Caccamise, chairperson of the Manual Language Department of the National Technical Institute for the Deaf, is responsible for research and instruction in manual/simultaneous communication and for instruction of deaf students. He is currently chairperson and director of the Project on Technical Sign Standardization and Development

xix

and is the author of numerous publications, including some on the efficiency of various modes of communication with hearing-impaired individuals.

Roger S. Fouts, associate professor of psychology at the University of Oklahoma, works in the fields of psycholinguistics, primate behavior, and comparative psychology. His publications in these areas include "The Use of Guidance in Teaching Sign Language to a Chimpanzee (Pan Troglodytes)" in the Journal of Comparative and Physiological Psychology *(1972) and "Communication with Chimpanzees" in E. Eibl-Eibesfeldt and G. Kurth (eds.),* Hominization and Verhalten *(Verlag, 1975).*

Nancy Frishberg, assistant professor of linguistics at Hampshire College, specializes in linguistic theory, the biological foundations of language, and sign languages of the deaf. She is a member of and certified by the Registry of Interpreters for the Deaf and the Sign Instructors Guidance Network (National Association of the Deaf), and has published several articles on sign language.

Timothy V. Gill is a systems consultant in the Information Services Department at Baxter-Travenol Labs, Inc., Deerfield, Illinois, where he works with real-time computer systems. He has published some 30 articles on primate language acquisition and primate learning, and he is a recipient of the Emory chapter of Sigma Xi's Annual Research Citation for his work on the LANA Language Project.

Gordon W. Hewes, professor of anthropology at the University of Colorado at Boulder, has conducted extensive research on human language origins, especially gestural theory, and on new models for hominization. He is the author of several publications on these topics and is the editor of Language Origins: A Bibliography *(2 vols; Mouton, 1975). Several professional societies have elected him a fellow, including the American Anthropological Association and the American Association for the Advancement of Science.*

Lyn W. Miles, assistant professor of anthropology at the University of Tennessee at Chattanooga and holder of a University of Chattanooga Foundation Professorship, has conducted research on language acquisition by apes and its implications for the origins and development of human language. She has published articles on language acquisition in apes and children and on the use of natural communications and sign language by chimpanzees, and has been awarded research grants from the University of Connecticut Research Foundation and the University of Chattanooga Foundation.

Francine Patterson *is a graduate student in the Department of Psychology at Stanford University, studying comparative developmental psychology. She has been conducting research with the gorilla "Koko" since 1972 and is the author
of "The Gestures of a Gorilla: Sign Language Acquisition in
Another Pongid Species" in D. Hamburg, S. Washburn and E.
McCown (eds.),* The Behavior of the Great Apes *(W. A. Benjamin,
in press).*

Duane M. Rumbaugh, *chairman of the Department of Psychol-
ogy and professor of psychology at Georgia State University,
is program coordinator for the Language Analogue Project
(LANA) at the Yerkes Primate Center in Atlanta and at the
Georgia Retardation Center. He is a member of the National
Institutes of Health Task Force on the Needs and Use for the
Chimpanzee as a Research Subject, a fellow of the American
Association for the Advancement of Science and the New York
Academy of Sciences, and a recipient of the Annual Research
Citation from the Emory chapter of the Society of Sigma Xi.
His publications include* Language Learning by a Chimpanzee:
The LANA Project *(with E. S. Savage-Rumbaugh and T. V. Gill;
Academic Press, 1977).*

E. Sue Savage-Rumbaugh, *a psychobiologist at the Yerkes
Regional Primate Research Center of Emory University in Atlanta, has published articles on the chimpanzee as an animal
model for language research, on play and sociosexual behavior
in captivity, on sexual dimorphism and on various aspects of
behavior in the pigmy chimpanzee, and on communication among
chimpanzees through learned behavior. She was a Woodrow
Wilson Fellow (1970-1971) and is a member of the International
Primatological Society, the Animal Behavior Society, and
other organizations.*

Acknowledgements

During the academic year 1976-1977, I had the opportunity of being affiliated with the Department of Anatomical
Sciences at the State University of New York, Buffalo (SUNYAB),
as Visiting Research Professor of Anatomical Sciences. The
opportunity was unique for a linguist like myself, in that it
was during this period that I acquired knowledge of gross
anatomy, neuroanatomy, histology, embryology, and neuro-
physiology through research and apprenticeship in the Department. I acknowledge with gratitude that the investment of a
portion of my career in basic sciences was a worthwhile
undertaking, in view of the fact that the accomplishments,
including the organization of two symposia, one with the
American Anthropological Association in November and the
other, with the American Association for the Advancement of
Science in February, were most fruitful.

While in Buffalo, despite the worst blizzard in 100
years, I received not only research facilities and instructions but also personal help and administrative assistance,
all of which contributed significantly to the completion of
this volume. I would, therefore, like to thank the faculty
and staff members of the Department of Anatomical Sciences at
SUNYAB for their valuable contributions. In particular, I am
greatly indebted to Dr. Russell Hayes for his professional
guidance in histology and assistance in slide-making, to Dr.
Rolf Flygare for sharing his research assets with me, and to
Dr. Harold Brody, Chairman of the Department, for his class
and laboratory instructions on neuroanatomy and for administrative support. However, I alone am responsible for any
errors that may be found in my work.

I am also especially grateful to the American Association for the Advancement of Science for the opportunity to
organize the symposium on which this volume is based. The
publication of such a volume is greatly appreciated. Thanks,

too, are due to Dr. Philleo Nash, Chairman of the Anthropology Section of AAAS, who made the symposium so much more enjoyable by providing the panel members with the needed transportation and by providing lodging for one of the panelists. It was through his aid that the symposium was made possible and was so well attended.

Finally, I wish to thank Dr. Harold Brody, Chairman of the Department of Anatomical Sciences, SUNYAB, Dr. Joseph C. Lee, Head of the Anatomical Sciences, the University of Oklahoma, formerly Director of the Gross Lab at SUNYAB, Dr. Robert L. Ketter, President of SUNYAB, and Dr. Albert Somit, Executive Vice-President of SUNYAB, through whose kind offices I was able to visit SUNYAB.

To Charles C. Thomas, The M.I.T. Press, Williams & Wilkins Co., and Consulting Psychologists Press, Inc. I wish to express my appreciation for permission to reproduce figures used in two of the articles in this volume. Thanks are also due to Dr. E. Lloyd DuBrul and Dr. Joseph S. Perkell for their kind consent to the reproduction of their figures.

Fred C. C. Peng
Tokyo, Japan
May 1977

Sign Language and Language Acquisition in Man and Ape

New Dimensions in
Comparative Pedolinguistics

Introduction

Fred C. C. Peng

For quite some time, a number of scholars in
such disciplines as psychology, anthropology, and
linguistics have sought to prove that nonhuman primates are indeed capable of language. They also
believe that a precursor of human communication systems in some definable form may be found among species other than Homo sapiens. Several attempts,
using oral language, the most famous of which was
the Viki Project by the Hayes, have been provided
in the past to show such a capability. All of them
have failed, a fact that is well-known in the history of primatology. (A similar project was repeated by a Japanese psychologist, Tsuneya Okano of
the University of Shizuoka, with a female chimpanzee by the name of Sachiko. But the project was
terminated after six months of training, when the
chimp barely learned to produce "mama," as the
trainer became convinced that there would be no
sense in continuing the project.)

Recently, however, starting with the work of
the Gardners, and followed by several other highly
successful projects in the United States, progress
is being made, which, using a different mode of
communication, has clearly demonstrated language-
like abilities in nonhuman primates and the probable
existence of a precursor of man's communication systems. More work obviously needs to be done, such
as the investigation of the communication systems
of nonhuman primates in their own natural habitat
(rather than in captivity) and of the extent to
which nonhuman primates can not only acquire a language of human origin but also transmit it to conspecifics.

This volume is an attempt to bring together
he latest research materials on sign language and
primatology. The approach adopted has been given
a new name: "Comparative Pedolinguistics." (The
term "pedolinguistics" was first proposed by the
late Karel Ohnesorg in 1972 as the result of the
First International Symposium for the Study of Child
Language held in Brno. It comes from Gr. <u>pais</u>,
<u>paidos</u> 'child', as in <u>pediatrics</u>, and <u>linguistics</u>.
The term has since become a familiar term in the
field of child language.)

In the present context, the newly coined term,
comparative pedolinguistics, is intended to mean
three things: (1) comparison of language acquisition
in different cultures, (2) comparison of language
acquisition across different modalities, and (3)
comparison of language acquisition between different
species. The first meaning, however, falls outside
the purview of the present volume which comes out
of the symposium entitled "An Account of the Visual
Mode: Man versus Ape," held in Denver, Colorado,
February 1977 under the auspices of American Association for the Advancement of Science. The revised
title, *Sign Language and Language Acquisition in
Man and Ape: New Dimensions in Comparative Pedolinguistics,* is thus meant to reflect more accurately the content of this anthology which attempts to
show that [1] sign language is a legitimate language
on a par with oral language and can be learned not
only by humans but by nonhuman primates as well,
and [2] nonhuman primates have the capability to
acquire a human language using a different mode.
(And perhaps, by making use of this mode, one can
trace some common traits regarding linguistic behavior between human and nonhuman primates.) Thus,
it can be said now that the language acquisition
ability is not species specific.

The contributions, greatly enhanced beyond the
symposium papers in scope and research orientation,
represent the works of anthropologists, psychologists,
linguists, and manual language experts. The contents are arranged so as to give a contrastive view
and a comparative outlook:

I.
    1. Sign Language and Culture (Peng)
    2. Code and Culture (Frishberg)

3. The American Sign Language Lexicon and
Guidelines for the Standardization and
Development of Technical Signs (Caccamise,
Blasdell, and Bradley)

II.

4. Language Acquisition in Apes and
Children (Miles)
5. Sign Language in Chimpanzees: Implications
of the Visual Mode and the Comparative
Approach (Fouts)
6. Language Skills, Cognition, and the Chimpanzee (Rumbaugh, Savage-Rumbaugh, and
Gill)
7. Linguistic Capabilities of a Lowland
Gorilla (Patterson)

III.

8. Linguistic Potentials of Nonhuman
Primates (Peng)
9. Comments and Remarks (Hewes)

In the first three papers of the volume, sign
language is put in its proper cultural context, so
as to allow for an assessment of its social status
in its historical and current perspectives. Specifically, then, sign language is also seen, not
as a self-contained entity, but rather as a general
norm of individual sign activities that are under-
going rapid development and changes in any given
society. Moreover, sign language, being a system
in the visual mode and of the "brachial apparatus"
(1), is considered to be a very useful tool which
can measure the similarities and differences of
certain aspects of linguistic ability — typically
in the realms of vocabulary, sentence construction,
and discourse — between man and ape. Thus, this
part of the present volume adheres closely to the
first purpose mentioned above; there should be no
doubt of the importance of sign language descriptions when dealing with language acquisition in apes.

In particular, Peng's "Sign Language and Culture" sets the tone for the volume. He points out
that the only language connection which we can show
at this moment between humans and nonhuman primates
is through the visual mode and that sign language,
be it American Sign Language (ASL) or Japanese Sign
Language (JSL) or any other, can serve this purpose
well, even though only ASL has been tried on a large
scale thus far (2). He also discusses in some detail (1) an historical account of language and culture, (2) erroneous views about sign language, (3)

characteristics of sign language, and (4) the deaf community in Japan.

Frishberg's contribution shows that even though there are similar signs which may be termed universals, they are combined in ways that are culturally determined; individual signs do not make up a language; they must be combined into phrases, sentences, and discourse units. When one takes these into consideration, it is difficult to say that there are universals other than individual signs and that sign language can be understood universally; if it can be understood cross-culturally, it is because the interlocutors make heavy use of the context of situations in which signing takes place.

In this connection, it must be added that some of the erroneous views quoted in Peng's article and the claim for universality in sign language cited in Frishberg's are related. That is to say, those who think that sign language is a surrogate or a primitive system would believe that it is simple enough to be understood universally and vice versa. In this sense, Frishberg's contribution is germane to Peng's.

The direct link between Frishberg's paper and those in the second part is that most chimpanzees and the gorilla Koko have learned ASL and have produced sentences and discourse units that can be understood by people who share similar background, say, ASL signers, but not by others, like JSL signers. Her title, "Code and Culture," has a significant bearing on this point: even nonhuman primates could learn to share the same code and culture with one group of humans (ASL signers) but not with another group (JSL signers).

The article by Caccamise et al. is another instance of code and culture; it supplements Frishberg's paper and illustrates Peng's major point — that sign language is as much a part of culture as oral language. But it also deals with another important topic: the creation of new signs. Moreover, the article implies that while humans may worry about the lack of suitable signs when confronting new situations and particular subjects (thereby resorting to creating new signs), nonhuman primates, too, as exemplfied by Washoe's BIB and WATER BIRD and Koko's inventions, may exhibit concern about the lack of suitable signs when confronting strange objects and new situations.

The second part of the volume is made up of four papers which are concerned with the assessment of the first-hand research progress, whereby apes of three different species are compared with humans and with one another, using sign language or some other visual-manual method. The theoretical implication of such a comparison is the possible existence of a common precursor of human and nonhuman communication systems and, ultimately, the discovery of the origin of a general faculty (3).

The papers on nonhuman primates pertain as a whole to the second purpose of the Symposium stated above. Miles' paper leads the reader from humans to nonhumans. The chimpanzee studies (with more general presentations like Fouts' before more specific case studies like Rumbaugh et al.) precede the gorilla study by Patterson.

Miles' article is a survey of the current projects on nonhuman primate language acquisition, along with a succinct comparison between human and nonhuman learning of sign language, from various standpoints, based on reports and results from a number of investigations, including her own.

Fouts' paper addresses two notions; namely, a comparison of the visual mode with the auditory mode and a comparison between man and ape by way of these two modes. First he emphasizes the differences between the two species, while adhering to his interest in similarities, and then compares the visual mode with the auditory mode. Here, he is in complete agreement with the opinions expressed by the authors in the previous papers. Like Peng, he is critical of contemporary linguistic dogmas, because he does not believe that man's primary mode of communication is in the auditory channel and thinks that this assumption is incorrect.

The paper by Rumbaugh et al. takes up the hard question of cognition among chimpanzees and deals with it from the point of view of two case studies, the Lana project and the pygmy chimpanzee project. While the Lana project has been known for quite some time through Rumbaugh et al., Language Learning by a Chimpanzee: The Lana Project, the observation of three captive pygmy chimpanzees is a

recent attempt which shows a systematic pattern of
their kinesic behavior in a sexual setting. Rumbaugh and his colleagues successfully argue (1) that
language should not be viewed as contingent upon
speech; (2) that the requisites to language are not
uniquely human; (3) that apes are more "intelligent"
than they have been thought to be historically; (4)
that the mentality of apes is probably closer to
man's than has been allowed historically; and (5)
that there is great potential for studying language-
type behaviors in apes and possibly in other animals as well.

Patterson's paper about the gorilla Koko con-
vincingly shows that another species of great apes
is also capable of language, so much so that Koko
now gives her distant relatives, the chimps, some
good competition in language acquisition. In one
sense, Patterson's paper provides a more human-
like datum for comparison between man and ape; in
this vein, she has successfully demonstrated the
conditions needed to disprove the claim for lan-
guage-specific capacity hitherto considered by in-
natists to be a unique possession of Homo sapiens.

The third part contains the remaining two contributions by Peng and Hewes. While Peng's paper
offers by way of integration a set of observations
that supports and explains some of the controver-
sial views vis-à-vis the linguistic capabilities of
man and ape, Hewes' paper reviews and evaluates the
preceding contributions.

It is hoped that this volume, as a whole, will
be a significant landmark in our long quest for the
origins of language and an important step forward
for a systematic and scientific investigation of
sign language. If these contributions can lead to
still more serious studies that will break the
inertia that characterizes contemporary linguistics
(4), the purposes of the Symposium will have been
served.

Notes

1. It should be mentioned that although bra-
chium usually refers just to the arm, in the absence of a suitable term that includes both the arm
and the hand, "brachial" instead of "manual" was
chosen to represent both. Thus, the term "brachial
apparatus" runs parallel with "vocal apparatus"

which is a common term in phonetics. But note that vocal apparatus includes the lips, the teeth, the tongue, etc., while "vocal" (or "voice") literally refers to the vocal folds only.

<u>2</u>. If Okano had tried JSL, instead of spoken Japanese, he might have succeeded in teaching Sachiko how to sign, just like Washoe or Koko in ASL. (Cf. <u>Josei Jishin</u>, July, 1977, p. 64).

<u>3</u>. General faculty may be defined as the collective function of the neurons in the central and peripheral nervous systems.

<u>4</u>. For a detailed discussion along this line, see Yngve (1975) listed in <u>References Cited</u> of "Sign Language and Culture," Note <u>4</u>, this volume.

1

Sign Language and Culture

Fred C. C. Peng

Introduction

The purpose of this presentation is the description in some detail of sign language and the culture of the deaf, a subject that has rarely been dealt with in anthropology. Sign Language is of great interest to me, because it not only represents a new dimension (perhaps a renewed one) in man's linguistic capabilities but also serves as a very useful tool to measure the similarities and differences of certain aspects of linguistic ability between man and ape (see also Chapter 8 in this volume). To achieve this purpose, I will divide the presentation into five sections: (1) historical account of language and culture; (2) erroneous views about sign language; (3) characteristics of sign language; (4) the deaf community in Japan; and (5) conclusion.

To avoid any possible confusion and to facilitate discussion that will follow in the course of my presentation, I shall ask the reader to keep in mind that for the most part the term "language" is synonymous with "oral language," unless specified otherwise, and that wherever necessary the term "oral language" will be used in contradistinction with "sign language."

Historical Account
of Language and Culture

It seems to me that nowadays in anthropology, if not in linguistics as well, the proposition that language is a part of culture is taken for granted.

The term "culture" was in effect invented by Tylor (1874) for anthropology. Tylor's well-known definition of culture is: "that complex whole which includes knowledge, beliefs, art, morals, law, custom, and any other capabilities and habits acquired by man as a member of society."

Students of anthropology will also realize that since Tylor's time cultural and social anthropologists have gradually moved from a definition of culture, that describes it as a more or less haphazard collection of traits, to one which emphasizes pattern and configuration. Kluckhohn and Kelly thus expressed this new concept of culture: "By culture we mean all those historically created designs for living, explicit and implicit, rational, irrational, and nonrational, which exist at any given time as potential guides for the behavior of men" (1945:97). Anthropologists have since Tylor used the term "culture" to cover everything in anthropology designated as archaeological, ethnographic, or even linguistic:

> "It is clear that language is a part of culture; it is one of the many 'capabilities acquired by man as a member of society'" (Hoijer 1948:335).

> "Language may no longer be conceived as something entirely distinct from other cultural systems but must rather be viewed as part of the whole and functionally related to it" (Hoijer 1953:554).

However, Voegelin dissents:

> "Some writers jump... to either or both of the following conclusions: (1) that language is a part of culture, which is debatable; (2) that the techniques for analysis of language and culture are the same or closely similar — this is surely an error (Voegelin 1949a:36). It may well be that anthropologists who work exclusively in culture, and linguists innocent of culture theory espouse this view, if they think of it at all; those who attempt to work in both fields seem more apt to appreciate the distinction and to speak of language <u>and</u> culture rather than of language as a part of culture" (Voegelin 1949a:36 fn. 2).

The cultural anthropologist Opler replies:

> "In fact, since I consider language (as do most linguists) to be a form of cultural activity and Voegelin does not, I suspect that I place even more

importance on what informants say than he does" (Opler 1949:43).

To this Voegelin answers:

"If language can be best comprehended as being a part of culture (which it is, according to Tylor and those who follow his definition of culture, including Opler), then why is it necessary to identify a critic of a particular theory of culture as a linguist? Is this not another example of subscribing to a definition formally and then forgetting about the definition in actual discussion? If language were merely a part of culture, then linguists should be competent to discuss other parts of culture by virtue of their training in linguistics. We must admit that if a linguist can discuss problems in culture, it is by virtue of his being a student in culture, also, rather than by a transfer from linguistic training; and vice versa" (Voegelin 1949a:45).

Both sides have something to say. It is true that language is a part of culture, especially when we recognize that "language" is only the general norm of human speech activities, as was so wisely put forth by Malinowski a few decades ago (Peng 1975a:6); but it is also true that linguists who have not been trained in the theory of culture in general cannot be expected to be completely conversant with culture theory.

The argument of whether or not language is a part of culture may thus be said to stem from two sources: (1) a lack of agreement on what culture is; and (2) a tendency to draw a strict parallel between a methodology and the substance to which such a methodology is applied.

In the first case, it seems clear that Voegelin and Opler did not agree on the definition of culture, although they might quote or refer to the same definition. This is shown in the following passage from Voegelin:

"If language were merely a part of culture, primates should be able to learn parts of human language as they actually do learn parts of human culture when prodded by Primatologists. No sub-human animal ever learns any part of human language — not even parrots. The fact that <u>Polly wants a cracker</u> is not taken by the parrot as part of a language is shown by the refusal

of the bird to use part of the utterance as a frame
(Polly wants a ...) with substitutions in the frame.
(For the three dots, a speaker of a language would be
able to say cracker or nut or banana or anything else
wanted). As George Herzog has phrased this, imitative
utterances of sub-human animals are limited to one
morpheme; to the parrot, then, Polly wants a cracker
is an unchangeable unit. From this point of view, we
can generalize: an inescapable feature of all natural
human languages is that they are capable of multi-
morpheme utterances (1949:45).

Obviously, Voegelin's statement here implies
that language has something that culture cannot
cover, a fact that has been known for ages about
language which may be called language creativity
(i.e., producing new utterances by way of putting
individual words together or of such transformations
as deletion, substitution, and addition), although
his illustration of substitution is only a small
part of it. (I must quickly add that we should not
lose sight of the fact that it is man, not "language," who has this kind of creativity.) Voegelin
suggested that "To discuss this problem calls for
a meeting of minds from both the culturist and linguistic camps" (1949b:45) (1).

The question of substitutability (or creavity
in its broader sense defined above), however, is not
limited to language, especially when we adhere to
the point that it is man who exhibits creativity
(and so do nonhuman primates). It follows that
music or other form of art such as painting (which
is also a part of culture) displays a similar feature. For instance, a jazz pianist can improvise
passages and change them at will, as he plays the
piano. And no one would deny that, according to
Tylor's, Kluckhohn's, or even Linton's definition
of culture (2), jazz music is a part of American
culture.

In the second case, it seems to me that what
Voegelin was saying is that if language is a part
of culture, then the techniques for analysis of
both must be the same or at least sufficiently similar, but since the techniques for analysis are
not at all the same, irrespective of erroneous
claims by some writers, language is therefore not
a part of culture. The other side of the coin is
that language is a part of culture, but since linguists use a different technique for its analysis,

linguists are not culturalists and, therefore, cannot be expected to know much about culture or the problems related to it.

This attitude pointed out by Voegelin was shared by some prominent anthropologists at that time, as may be evidenced by Linton's reasoning for excluding linguistics from the scope and aims of anthropology:

> "Turning to the field of cultural anthropology and its subsciences, we find that the subscience of linguistics is, at present, the most isolated and self-contained. The study of languages can be and largely has been carried on with little relation to other aspects of human activity ... That linguistics ultimately will be of great value for the understanding of human behavior and especially of human thought processes can hardly be doubted. However, work along these lines has barely begun and linguistics is still unable to make any great contribution toward the solution of our current problems. For that reason it has been ignored in the present volume" (1945b:7-8).

The controversy here can hardly be exaggerated. But, I am of the opinion that if language is a part of culture, even though the techniques employed for its analysis are different, the sheer difference in technique alone does not automatically give rise to the assumption that linguistics is totally unable to contribute to the solution of problems in anthropology, so long as the substance to which the techniques are applied is not tampered with and so long as claims for each technique do not exceed the success it can achieve (3).

On the other hand, however, there is some truth in the opinion expressed by Linton about linguistics, a fact that is being painfully experienced now more than thirty years later by many contemporary linguists, simply because linguistics has moved farther and farther away from its relevance to man and his environments, so much so that it can no longer be recognized in some circles as linguistics (4). I should also point out that en route to alienation from the proper domain of language and culture, there was and may still be a time during which certain practitioners (those from MIT in particular) preferred to call their linguistics a branch of psychology and that still others so practice their linguistics as to make it

devoid of human interaction which, because the substance has been tampered with, is better called mathematical linguistics, symbolic logic, or comput-
ational linguistics, rather than linguistics per se
(Peng 1975a).

As early as 1951, linguists (then structuralists) were already in the process of divorcing linguistics from anthropology, albeit with mixed feelings, as may be seen in an interesting statement
by Voegelin himself:

> "The new secretary of the Linguistic Society of
> America would define linguistics not merely as a
> branch of anthropology but, more specifically, as a
> branch of 'cultural anthropology'; another linguist,
> cited in the same article, speaks of linguistics as a
> branch of, or rather, a kind of mathematics" (1951:364).

I should add that when the MIT group came up
with something that really looked like mathematics
a few years later, the structuralists were caught
quite unprepared. Since the time of Neogrammarians,
linguists have had a tendency to claim more for
their methodology than it may deserve. However, I
do not wish to suggest that mathematical techniques
or some other tools must not be used in dealing with
language as a part of culture or that if such a
technique is employed to handle language, language
is no longer a part of culture, a proposition that
was precisely the source of argument stated earlier.
What I am saying is that so long as the substance
with which we are concerned is not tampered with
(in our case, if language is not taken out of its
human and environmental contexts), any method or
technique can be used as long as the technique is
not elevated into a kind of intellectual fetish.

Perhaps an example is needed to illustrate
the essence of my word of caution. For instance,
the transformational-generative paradigm, which
flourished in the 1960s but has since declined, is
a data-processing technique. But its proponents,
encouraged by the tendency of over-claim in linguistics, claimed it explained or accounted for
human mental activity; it was even said to characterize language acquisition stages or the ideal
hearer-speaker's ability that is supposed to under-
lie some kind of language universals, even though
there is no empirical support nor evidence for its
feasibility. The following is a case in point of

this over-claim that prevailed throughout the 1960s:

> "... It is now proposed that, first, children are
> born with a biologically based, innate capacity for
> language acquisition; secondly, the best guess as to
> the nature of the innate capacity is that it takes the
> form of linguistic universals; thirdly, the best guess
> as to the nature of linguistic universals is that they
> consist of what are currently the basic notions of
> Chomskian transformational grammar. Metaphorically
> speaking, a child is now born with a copy of <u>Aspects of
> the Theory of Syntax</u> tucked away somewhere inside.
> Given the present state of knowledge regarding innate
> capacity and language universals, the above seem de-
> fensible guesses" (<u>5</u>) (Fraser 1966: 116, paraphrase of
> McNeill 1966).

My whole point here (cf. Peng 1975b and 1976a)
is that there is an urgent need to bring linguistics
back to anthropology and treat language in its broad
sense (i.e., not limited to oral language) within
the bounds of culture, whereby man and his environment in connection with his linguistic behavior
should once again become the focus of attention.
It is a good thing that the tide is changing in this
direction, as may be evidenced by recent activities
in sociolinguistics, pedolinguistics, and neurolinguistics.

Some Erroneous Views
About Sign Language

Erroneous views about sign language have come
not so much from laymen but from professional linguists who stated these views in the name of science.
The most serious error may stem from ignorance about
sign language and conceit about oral language. This
attitude is extended to nonhuman primates, when man
compares himself with ape: "For a variety of reasons, certainly including ignorance and possibly
conceit," say Eimerl and DeVore, "man has always
had a tendency to consider his own qualities as
being unique" (1974:16).

To substantiate this, a more general argument
between de Saussure and Whitney should be mentioned
(cf. Peng 1975b:13). In this dispute, de Saussure,
for all his contributions to linguistics, took, in
my opinion, a wrong turn regarding <u>les image visuel-
les</u> and dismisssed too quickly Whitney's view as
<u>trop absolue</u> (too extreme). Here is what he said:

"Ainsi pour Whitney, qui assimile la langue à une
institution sociale au même titre que toutes les autres,
c'est par hasard, pour de simples raisons de commodité,
que noun noun servons de l'appareil vocal comme instrument de la langue: les hommes auraient pu aussi bien
choisir le geste et employer des images visuelles au
lieu d'images acoustiques. San doute cette thèse est
trop absolue; la langue n'est pas une institution sociale en tous points semblables aux autres ...; de
plus, Whitney va trop loin quand il dit que notre choix
est tombé par hasard sur les organes vocaux; il nous
étaient bien en quelque sorte imposés par la nature"
(1955:26).

More recently, however, Voegelin and Harris
said of sign language:

"Sign language of the plains or Australian type is
not derived from any one language any more than is the
Chinese ideographic writing ... derived from one of
the Chinese languages, or any more than are Arabic
numerals derived from a spoken language. But alpha-
betic writing is derived from the one language of
which it is the 'written form'; and the deaf and dumb
sign language is in turn derived from a given 'written
form' of a given language" (1945:459).

It is of interest to note that Voegelin and Harris
make a distinction (albeit a fallacious one) between
sign language of the plains or Australian type and
sign language of the deaf; they are right about the
former but wrong about the latter which is confused
with fingerspelling.

A similar erroneous view regarding sign language of the deaf was expressed by Hill:

"Even the manual language of the deaf is derived
from the pre-existent spoken language of the community" (Hill 1971:25).

Most recently, a surprising prejudice against
sign language was expressed by Bender (6):

"Spoken language is primary; sign languages are
surrogates ... A lot of work on sign language has not
come up with a substitute for the fantastic efficiency
achieved by the coding mechanisms used in speech. Two
obvious speech surrogates — shorthand transcription
and reading of printed text — are more efficient
than sign language" (1976:20).

It would, of course, be misleading to argue that only linguists hold erroneous views about sign language. The truth of the matter is that sign language in every human society is looked down upon and its cultural value is greatly distorted by most hearing people. In Japan, for example, even casual observations of language attitudes make it immediately apparent that the language of the hearing is felt to be far superior to Japanese Sign Language (JSL). JSL has no prestige and is not even granted status as a proper language by most hearing people (Peng and Clouse 1977).

But why do people feel this way about sign language? Several reasons may come to mind, such as majority versus minority, when one compares the sheer sizes of the hearing and deaf communities, or hearing ability versus hearing impairment, when competition is of major concern. One very important reason seems to summarize all the rest: Oral language has been regarded as the origin of language in man's history, with a theological tint in the West, so much so that an alternative based on the use of a different mode is utterly unthinkable. There is a fear that any such possibility would downgrade man's dignity and cast doubt upon his unique qualities. Behind this fear lies a very real superiority complex, as seen in the following statements:

"Clark Hull once suggested that since infrahuman primates failed to learn human language even after being prodded in this direction by primatologists, it might be theoretically interesting to try to teach chimpanzees in a laboratory a simple sign language to see whether or not, thus equipped, they might not exceed all previous scores in operating complex 'chipomats' or other machines comparable to our gymnasium lockers in which we have to remember a series of turn-left-turn-right combinations" (Voegelin 1951:368) (7).

"The kind of simplified language which Clark Hull recommended for teaching to apes had this characteristic of human languages: it consisted of a system of discrete units in which one or more units might be held constant while one or more units varied; these units (or morphemes, as we say in linguistic analysis) might be produced by hands and feet rather than by lungs, larynx, mouth, and nose. The last named organs are not very different in humans and in chimpanzees; in fact, the lips of chimpanzees are more extensible than those

of man, and should therefore function exceptionally
well in producing rounded vowels ... What one primatological teacher seems not to comprehend, however, is
that his single-morpheme-uttering chimpanzee still
falls short of having any language at all, regardless
of whether other associations are made with the utterance in question (such as reaching for a cup). The
criterion for a language is that there be variability
within a fixed frame. Subhuman utterances, however
long they may be, are functionally single morphemes.
In contrast, every human language shows variability
within fixed frames" (Voegelin 1951:369).

"Human language and primate communication (whether
with hand signals or plastic shapes, etc.) may be seen
as differing in degree only, but the difference is
crucial. In fact, the difference in complexity in
terms of phonological and syntactic encoding between
lower primate communication and human language is a
gulf ... Even if we leave aside the important matter
of development of adequate vocal apparatus, lower primates do not have the neurological mechanisms and cognitive capability for language" (Bender 1976:20).

The above passages show that many linguists,
with few exceptions, have great difficulty accept-
ing sign language as a part of or a form of language on an equal footing with oral language.
They have greater difficulty with the fact that
sign language is a part of culture in the way oral
language is. But to pursue the line of research
on the origin of language, a search that has been
"forbidden" since the moratorium of the Société
Linguistique de Paris one hundred years ago but was
"reopened" in New York in 1975 (8), necessitates
the acceptance of sign language as a legitimate
form of language on a par with oral language. There
are two reasons for this: (1) there is a definite
limitation to what historical/comparative linguistics and archaeology can do to reveal traces of the
origin of language; and (2) the study of nonhuman
primates must be brought into the comparison with
humans vis-à-vis their respective communication
systems, in which case, the only language connection
that we can show at this moment between humans and
nonhuman primates is in the visual mode.

Characteristics of Sign Language

Six characteristics, based on Japanese Sign
Language for the most part, may be singled out here

B) members are interested ... keeping
...ng with a member of the Health Professions
... an additional advocate in compiling your cover
...ber will try to get to know you. This meeting
... interview so use it as a learning experience.
...s as though you were going to a medical school
... member once. For questions about choosing
..., return to your Pre-Med Advisor at the CPPS

you meet with your pre-med advisor. Realize
... you to someone whom you know. If there is a
..., ask him/her write you an individual letter of
... appointment with the Pre-Med Advisor, you
...le an appointment. Remember that the HPAB
... their students the week after MCAT's!

for a cursory discussion. They are: (1) Simultaneity; (2) Reversibility; (3) Directionality; (4)
Place Names; (5) Sign Symbolism; and (6) Reduplication and Repetition.

Simultaneity

The term simultaneity is not new. But I must
point out at the outset that I am using it in a
quite drastically different sense from, say, Frishberg (1975). Frishberg's use of the term seems to
prevail in the United States (9):

> "A sign can be specified by describing its four
> simultaneously realized parameters: hand configuration,
> location, movement, and orientation. The first three
> of these were suggested by Stokoe, Casterline and
> Croneberg 1965, and were termed by them 'dez', 'tab',
> and 'sig' respectively. These three will separate
> minimal pairs such as APPLE:CANDY (differing only in
> hand configuration), APPLE:ONION (differing only in
> location) and APPLE:CHEEKY (differing only in move-
> ment). The last parameter, orientation, has been
> found necessary to distinguish pairs like SIT:NAME
> or SHORT:TRAIN, which differ only in the angle at
> which the two hands contact one another. Let me em-
> phasize again that these four aspects of a single sign
> are articulated at once; the hand or hands assume a
> configuration at a particular location, in a specific
> orientation with respect to the body, and execute some
> distinctive movement. It is true that this movement
> dimension will continue over time (albeit relatively
> short in duration), and one sign will follow another
> in sequence; but the level of analysis within a single
> sign is not segmental along a time axis" (1975:698).

To account for the difference in usage, let me
add that the kind of simultaneity mentioned by Frishberg and others is a common phenomenon observable in
any oral language, whereas the kind of simultaneity
I have in mind is unique to sign language, such as
JSL, and cannot be found in any oral language at all.
Frishberg's notion of simultaneity is commonly observable, say, in English, because it is articulatory. Mine is unique, because it is conceptual (perhaps cognitive would be a better term). To contrast
my notion with hers, I should quote my statement:

> "... the equivalent of otoko ga onna o naguru 'a
> man hits a woman' is man (hit + woman) where hit and
> woman within the parentheses are signed together. By

'together' I mean literally that this is really what happens. But note that in either Japanese or English there is no way to express this simultaneity, because each is bound by time and the order of the elements differs; that is, in time sequence Japanese expresses onna 'woman' before naguru 'hit' whereas English expresses hit before woman" (Peng 1975b:15).

It is clear that what Frishberg means by si-multaneity is the simultaneous articulation of the four parameters in one sign, whereas my simultaneity has to mean the co-articulation of two signs. In discourse, however, we may also observe a different kind of simultaneity. For instance, if a Japanese is asked "How many brothers do you have?" he is ex-pected to first give a number, say, 3, meaning two brothers and himself, and then he is asked another question like "Are you older or younger?" to which his answer can be, say, "the youngest." If the same conversation is conducted in JSL, the situation becomes quite different. After the initial inquiry of the number of brothers, a deaf person will in-variably sign "three, with 3 fingers (index, middle finger, and ring finger) of one hand and at the same time using the index of the other hand to either hit or touch the last of the three fingers stretched out, if he is the youngest," thereby cutting the conversation to half the length of its oral counterpart, making the two oral utterances "Are you older or younger?" and "The youngest" un-necessary.

While the two kinds of simultaneity I have described are absent from any oral language, the kind of simultaneity Frishberg has mentioned exists in every language, oral or manual. Note that what she has referred to as dez, tab, and sig corres-pond neatly to articulator, point of articulation, and manner of articulation, respectively, in pho-netic terms, if and only if dez is not articulated away from tab. In this sense, then, a sound like [d] may be said to have simultaneity, because its dez (the tip of the tongue), tab (the alveolar ridge), and sig (stopness) must be articulated at once; a closer resemblance is an affricate where sig (the manner of articulation) constitutes a movement (from stop to fricative). The only dif-ference is notational (10).

Notational aside, we must now return to the characteristic of simultaneity in JSL. Although I

have mentioned two examples, one for each kind of
simultaneity, I have come across many more examples,
such as "two people go (some place)" where TWO
PEOPLE + GO are signed together (with two vertical
index fingers moving forward) and "(a group) is
split into two" where SPLIT + TWO are signed together (by placing the interior edge of the right
hand [Peng 1976c:189] between the left index and
middle fingers). It is important, therefore, to
determine the extent to which the characteristic of
simultaneity is used in JSL (and other sign languages around the world) and to classify those signs
that involve simultaneity. A clear understanding of
simultaneity in sign language should shed light on
the eventual analysis of not only JSL but other
sign langauges as well, for simultaneity cannot be
analyzed linearly, such as A $\longrightarrow$ B + C, a formula
based on oral language that is inadequate, to say
the least, to handle JSL (or any other sign language
for that matter). The task of analysis is complicated by the fact that simultaneity is often accompanied by directionality (see the discussion below).

Reversibility

Reversibility is another characteristic of JSL
that cannot exist in any oral language. Reversibility is commonly observable in antonyms; for instance, good/bad and bright/dark. By reversibility
is meant that if two signs have the opposite meanings,
their hand movements are the reverse of each other.
For example, GOOD (in the sense of skillful and dex-
terous) is signed with "right palm going down on
left arm" while BAD (in the sense of clumsy) is
signed with "right palm going up on left arm." Likewise, BRIGHT is signed with "open hands, palms facing
outward, from crossed position to wide open position"
but DARK is the opposite when the open hands are
crossed before the face. There are many more signs
that show this characteristic, e.g., CLEVER:STUPID,
FAR:NEAR, and HIGH:LOW.

In oral language, some kind of reversibility
(in terms of articulation) can be found, e.g., yes
[yɛs] and say [sɛy] but they are not antonyms. In
American Sign Language (ASL), BRIGHT:DARK come
rather close to reversibility but are not quite the
same: BRIGHT (both "AND" hands point forward, index
tips touching; open the fingers as the hands are
moved upward and to the sides ending in a "5" position, palms facing forward) and DARK (the open hands,

palms facing self and pointing up, are crossed before
the face).

If all antonyms in JSL showed the characteristic of reversibility, the research ahead would be
quite simple. The problem is that not all antonyms
in JSL, due to historical reasons, show this characteristic, e.g., TRUE:FALSE, OLD:YOUNG, HARD:SOFT,
and EASY:DIFFICULT. The specific aim in this kind
of research is to document all antonyms in JSL (or
any other sign language that shows this characteristic) and classify them according to the characteristic of reversibility. But the task is complicated
by the fact that there are varieties in reversibility and that reversibility may be compounded by
directionality. Two may be mentioned here for discussion.

In the case of BRIGHT:DARK, the hand configurations do not change; only the movement is changed
(i.e., reversed). In the case of KNOW:NOT KNOW,
however, reversibility is present but tab is shifted
to a different location. To wit: KNOW (right palm
caressing down the center of the chest) but NOT KNOW
(right palm caressing up on the right breast). All
varieties of the characteristic of reversibility
must, therefore, be exhausted, in order to work out
a sensible taxonomy. When such a taxonomy is made
available, it will shed light on and perhaps even
provide a new insight about antonymity which has
been taken for granted in oral language and whose
definition has never been seriously questioned. For
instance, are "forget" and "remember" or "marriage"
and "divorce" antonyms or not in English? In terms
of reversibility, they are antonyms in JSL.

The reason for this kind of incompatability is
that oral language has only one criterion, viz.,
meaning, to work with, whereas sign language has at
least two criteria, viz., form and meaning, on which
to account for and test antonymity. Moreover, antonyms are often thought of as adjectives rather
than verbs or nouns, a misconception that should be
rectified.

Directionality

Directionality is another characteristic of JSL
that is absent from any oral language but also exists
in ASL. This characteristic has to do with the relative positions the signer and the receiver take,

when they constitute a dyad in a conversation, and/
or with the protagonists' relative position to each
other and to the signer in the conversation or in a
story during the conversation. Examples are SAME
and DIFFERENT.

Note that these two signs are antonyms in terms
of meaning but not in terms of form (for they do not
show the characteristic of reversibility). In contrast, however, they show the characteristic of directionality. Normally, SAME is signed with "the
hands one foot apart, palms up and parallel with
the body, while pinching the thumb and the index
together with the remaining fingers folded." But
when the signer and the receiver are facing each
other and the topic of conversation is concerned
with them, such that an agreement between them must
be reached, SAME is, then, signed with the line
between the two hands perpendicular to the body,
other things being equal. DIFFERENT is similarly
signed, except that the thumb and the index of each
hand form an "L" shape and, instead of pinching, the
right hand turns clockwise, while the left hand
counter clockwise. The characteristic of directionality comes into being just like SAME, when the
signer and the receiver are facing each other and
their opinions must now differ. Other examples are:
LIMIT (ceiling versus boundary versus bottom), TELE-
PHONING (going out to the right or to the left),
CONVERSATION (i.e., TALKING), and MEET (i.e., to
meet someone). Directionality adds a fine-grained
connotation to a conversation in JSL that must be
laboriously described and explained in any oral language. Thus, in terms of functional load, a sign
with directionality definitely carries much more
information than its equivalent in an oral language
(see also Frishberg's examples in this volume).
More research is badly needed in this area to determine the impact of directionality on sign language
communication. At present, I have no way of measuring this impact other than documenting all signs in
JSL and any other sign language that show this characteristic and cross-reference them with simultaneity and reversibility.

Place Names

One of the richest areas for research in JSL
lies in place names (see "Place Names in Japanese
Sign Language" [Peng and Clouse 1977]). While the
study of place names in spoken and written languages

has been highly developed in Europe, my work on this topic may be the first in any sign language. More-over, findings based on this research have already led to the construction of what may be called the "Folk Etymology" of JSL place names, a contribution to the knowledge of the history of JSL that is other-wise difficult to obtain.

More often than not, the term "Folk Etymology" is reserved for <u>erroneous</u> etymologies involving some sort of remodeling of a less frequent or favored pat-tern into a more favored one:

> "Possibly of greater amusement than significance in the development of languages are new formations that represent fanciful modification, such as English <u>sir-loin</u>. This is from Fr. <u>sur-loin</u>, in which the first element derives from Lat. <u>super</u> 'upper', so that his-torically the word refers to the upper part of the loin. In English, however, <u>sur</u>, which was not found in other widespread compounds, seemed aberrant and was modified to the apparently sensible <u>sirloin</u>, for the upper part of the loin as a noble piece of meat. Somewhat scorn-fully, this process has been referred to as Folk Ety-mology" (Lehmann 1973:193-4).

In this presentation, however, the term should not be understood to imply that the explanation for the origin of a particular sign is necessarily cor-rupt. Rather, it is intended to indicate that the explanations for the derivations of place names are <u>a linguistically expressed reflection of the cog-nition of contemporary JSL native signers</u>, a part of their culture, so to speak. In this sense, as has been admitted by Lehman, Folk Etymology does not differ essentially from the process by which a con-temporary English suffix -<u>burger</u> was formed, or an Old English suffix -<u>ling</u>, or an Old High German suf-fix -<u>er</u>.

As to the question of whether these present-day explanations represent authentic derivations, con-clusive verification is difficult owing to the fact that JSL, like other sign languages, is without a written form, which precludes the use of historical documents for attestation and for conducting dia-chronic comparisons. Verification can be made only in cases of borrowing where the source of the sign can be traced back to its foreign model. At the same time, however, linguistic forms cast in visual images characteristically exhibit a degree of

iconicity rarely observed in the predominantly arbitrary acoustic images of oral languages, thereby
leading one to believe that explanations forwarded
by signers may be quite true to historical fact, if
the following two conditions are met: (1) the trans-
parency or credibility of the derivation weighed in
terms of the cultural background of native signers;
(2) the agreement between signers in their independently given explanations of the same place name
sign and its derivation.

Admittedly, these conditions alone do not in-
sure that a sign's suggested etymology is completely
reliable, but they do provide a rough rule of thumb
for discerning which derivations are the most ques-
tionable. It is of interest to note that when in-
terviewing informants, it was found that for the
most part signers could readily supply the reasons
for the use of a sign as a specific place name with
the explanations offered by different informants on
separate occasions generally supporting each other.

When introducing new place names into JSL, two
alternatives are open to signers. They can either
bring in the place name from another language (usually Japanese) through borrowing or coin a sign for
the locality independently by making use of the
native JSL system.

The actual distribution of place name signs
enables the construction of what can be called "The
Scale of Adaptability" according to which one can
gain some idea of which foreign terms are most likely to be borrowed into JSL and which are not. On
this scale, Japanese terms that can be readily analyzed into their constituent morphemes are said to
stand high on the scale, whereas those terms that
are unanalyzable or only analyzed with great difficulty are said to be low on the scale. Thus, it can
be stated that terms showing a high level of adaptability will most probably be adopted from Japanese
and integrated as loanshifts or loanblends, while
those with a low adaptability level will most likely be domestically coined.

The scale just explained is only one of the
findings that the study of JSL place names has produced. Another one may be mentioned in passing,
that is, "A Relative Chronology of JSL Place Names
Sign Types." Although it is impossible to give a
specific date for the inception of each place name

sign, because of the absence of any JSL written notation or recorded history, it is possible to establish a chronology of the sign types in relation
to each other. More discussion along this line is
available in Peng and Clouse (1977).

Sign Symbolism

The term "sign symbolism" is new; it is coined
after the model of sound symbolism. Whether or not
the two are on a par with each other is immaterial
here. The main point is that the Japanese deaf
divide the human body into several segments and at-
tach some culturally defined meaning to each such
segment, so much so that those signs that share the
same body part seem to bear resemblances to that
meaning (11).

It is anthropologically well-known that each
culture has its way of cutting up time and space in
relation to the human body. The Japanese deaf regard the front as the future and the back as the
past, a fact that is in sharp contrast to the Aymara
Indians of Peru who place the future in back (because one cannot see it) and the past, in the front
(because one can see it). Accordingly, then, many
signs in JSL that have the forward movement carry
the meaning of "future," while those with the backward movement carry the meaning of "past." For instance, NEXT YEAR, TOMORROW, NEXT WEEK, FUTURE, etc.,
have the forward movement but LAST YEAR, YESTERDAY,
LAST WEEK, PAST, etc., have the backward movement,
with the body as the dividing line. In both cases,
the term "sign symbolism" is applied.

The Japanese deaf also regard the head as the
center of thinking and, consequently, such signs as
THINK, DREAM, CLEVER, STUPID, MEMORIZE, FORGET, WOR-
RY, STUBBORN, and SCRUTINIZE all pertain to the head.
By the same token, the Japanese deaf think that the
arms are relevant to power. Thus, SKILL, POWER,
EXPERT, GOOD (in the sense of good at), BAD (in the
sense of clumsy), METHOD, and LABOR (i.e., LABORIOUS)
all pertain to at least one arm.

ASL seems to show some sign symbolism but not
to the extent of JSL. For instance, according to
Lottie Riekehof's Talk to the Deaf (1963:18-26),
many signs such as THINK, MIND, KNOW, INFORM, RE-
MEMBER, and FORGET pertain to the head at some point
(12). If this is the case, it will be of great

interest to look into the sign language behavior of
the deaf throughout the world to see if there is any
universality involved. This task, of course, must
await the completion of a thorough investigation of
at least a good number of different sign languages
and, therefore, does not concern us here. The intriguing problem of sign symbolism in JSL, however,
remains as a vital part of this presentation on sign
language and culture.

Reduplication and Repetition

Reduplication and repetiton are also common in
many oral languages. Examples are: <u>so so</u>, <u>very</u> very
<u>much</u>, <u>willy-nilly</u>; <u>bochi-bochi</u> 'gradually, about
time', <u>chobi-chobi</u> 'a little at a time', <u>toki-doki</u>
'at times' (in Japanese); and <u>ch'ang-ch'ang</u> 'always',
<u>kuei-kuei</u> chü-chü 'well behaved', <u>huan-ying</u> huan-
<u>ying</u> 'quite welcome' (in Chinese). The distinction
between reduplication and repetition is not clear,
as far as I can tell from the literature.

These characteristics have a far-reaching theoretical implication for the analysis of sign language. In a very interesting paper presented by T.
Supalla, himself a deaf signer, entitled "Systems
of Modulating Nouns and Verbs in ASL" (n.d.), he
reported that repetition may be used as a functional
criterion to separate a noun from a verb when their
hand configurations are the same. The examples he
gave are SIT versus CHAIR. According to Dean A.
Christopher (1976:213), these two signs have different hand configurations. But in Supalla's case,
the two have identical hand configuration; that is,
CHAIR is signed with two "H" hands placed crosswise,
except that the top "H" hand is bent. The distinction, then, depends entirely on repetition: In the
case of SIT, no repetition is required; in the case
of CHAIR, the top "H" hand taps the bottom "H" hand
twice as a repetition. He concluded tentatively
that in citation forms, if hand configurations are
the same, nouns exhibit repetition but verbs do not.

In JSL, the situation is just the opposite. I
have found that verbs usually have repetition, especially when used in the imperative, but nouns do
not. A good example is GET MARRIED versus MARRIAGE.
They are signed with the right thumb and the left
little finger coming together (see the variations
in Peng n.d.a, for the illustration of directionality). But in the case of GET MARRIED, the thumb

and the little finger tap each other twice, while
in the case of MARRIAGE no repetition after the
contact is made. And, interestingly enough, by
sheer coincidence or what historical linguists would
call "chance resemblance," in JSL, SIT and CHAIR
are signed with the same hand configurations as in
ASJ. But since the functional criterion is the opposite, SIT has repetition and CHAIR does not.

With regard to reduplication, it differs from
repetition in that the recurrence of a sign at least
once is required; without the recurrence, the sign
itself either does not exist as a free form or has
a different meaning. Examples are: ALWAYS (lit.
'every day' which is signed with the hands one foot
apart and they move in an inward circle twice while
each hand holds an "L" shape), SOME TIMES (which is
signed with the right index from left to right making two semi-circles). In the case of ALWAYS, if
no reduplication is involved, the hand configuration
moving upward means EAST or THE SUN (thus, 'every
day' means the sun going up and down). But in the
case of SOME TIMES, if no reduplication is involved,
one semi-circle is not a free sign in JSL.

In contrast to this description and illustration of reduplication, repetition may be said to be
the recurrence of a sign at least once that is not
required; without the recurrence, the sign itself
can stand alone and the recurrence more or less adds
some emphatic connotation to the original meaning
of the sign (as in CHAIR and SIT) or is simply a
rhythmic habit of the signer (like hesitation) that
is taken for granted or attracts the attention of
the viewer. I must add that repetition as a functional criterion does not contradict the description
I have just offered. Because it refers to citation
forms only; in actual context, CHAIR is seldom mistaken as SIT, whether repetition takes place or not.
Other examples in JSL of repetition are: FUNNY (which
is signed with the right fist pounding the ipsi-
lateral abdomen twice or more), SKILL (which is
signed with the right index tapping the left forearm twice), and ALL RIGHT (which is signed with the
right little finger tip tapping the chin twice).

The Deaf Community in Japan

I hope I have convinced the reader that sign
language is a part of language, on a par with oral
language and that, consequently, sign language is a

part of culture. But sign language, as was already
amply illustrated, can in some instances be much
more sophisticated and revealing than oral language.
The reason may be in the fact that sign language is
more behavior-oriented in that extrinsic rather than
intrinsic body parts are used and that the brachial
apparatus consists of two upper extremities while
the vocal apparatus has only one tongue. It is because of this anatomical difference that no oral
language is capable of simultaneity, reversability
(i.e., reversing a sequence of sounds to make an
antonym), or directionality (i.e., placing a sequence of sounds in a slightly different position
to convey a fine-grained shade of meaning).

In order to have culture, however, the deaf in
any society must have a community of their own, although it need not be geographically bordered, such
as the black neighborhood, the Chinatown, or the
Polish community in the United States where cluster-
ing is prominent. Because it is through such a community, where human interaction takes place, that
sign language can be directly linked to culture.
Let me, therefore, turn to an ethnographic description of the deaf community in Japan to exemplify
this linkage.

There are approximately 230,000 to 250,000
deaf people in Japan. The approximation is due to
the fact that the figure represents only the re-
gistered deaf people, congenital or otherwise, and
that there are many hidden deaf persons whose existence is not known to the government or the All Japan
Federation for the Deaf (Zen Nippon Roa Renmei).
Efforts are being made constantly by the Federation
and its local chapters to locate these people. Thus,
there may be as many as 0.3% of the total population
who are deaf. This is a sizable minority in Japan
compared to other minority groups, such as the Ainu
whose number has been in the vicinity of 18,000 over
the past four hundred years (cf. Peng and Geiser
1977).

Of these quarter of a million or so deaf people
in Japan only 75% are considered literate or semi-
literate. The deaf, literate or not, are scattered
throughout the country in small groups, sometimes
a group of two or three within one household, re-
siding amid the hearing majority in various parts
of urban and rural areas. But in each prefecture,
there is always a local organization, a chapter of

the Federation, which functions as the center of
coordination to serve the deaf population within the
prefecture. It is through such an organization that
the deaf maintain their contact and communication
among themselves beyond their immediate neighborhood.
The national Federation, then, organizes an annual
meeting, usually in the summer, whereby deaf people
from all over the country can get together and share
their fellowship for a few days. During the conference, the Federation passes on information concerning welfare and other items such as government
business and the job market.

In each city or rural area, there is also a
building called Welfare Center for the Handicapped
People (Shinshin Shōgaisha Fukushi Kaikan) or its
equivalent where the deaf meet in small groups or
hold seminars, socialize, and discuss various problems of their interest practically every night.
Interested hearing people are occasionally invited
and may sometimes take part in such meetings on
their own. In the heart of Tōkyō, moreover, there
is also a coffee shop where the deaf invite their
friends, deaf or hearing, and relax in an atmosphere
that is quite pleasant. Usually, there are about
100 deaf people in the coffee shop on the weekday
evenings and perhaps more on weekends.

Needless to say, in every type of gathering,
national or local, business or pleasure, the means
of communication is JSL. Many, at least among the
young urban signers living and working in Tōkyō,
who have completed senior high school and in scattered cases, college or university, are bilinguals
with varying levels of ability in the Japanese of
the hearing majority and JSL. This gives rise to
a diglossic situation comparable to that described
by Stokoe (1972:154-67) in relation to American
signers. For Japanese signers, the High Language
is Japanese while the Low Language is JSL. Between
these two extremes there exists a continuum with
hybrid forms of JSL showing greater or lesser de-
grees of direct influence from the High Language.

Even a cursory look at the real situation reveals great complexity in sign language and sign
language behavior. Strictly speaking, even the convenient label, JSL, that has been used here so far,
is too loose, since it implies the existence of some
sort of single and uniform sign language employed
by the majority of signers throughout Japan. But

this is not the case. Presently, JSL is a collective term that encompasses a vast array of coexist-
ent varieties with no standard form, though attempts
at devising such a form are beginning to emerge.
There is an indeterminate number of dialects, both
geographical and social, which are further complicated by processes of cultural contact and diffusion,
whose effects are manifested in such sociolinguis-
tically defined phenomena as pidginization, diglos-
sia, and code-switching.

The contact between the deaf minority and the
hearing majority depends entirely on the bilingual
deaf signers and hearing interpreters (professional
or volunteers). For those deaf signers who have
not learned lip-reading or oralism, communication
with any hearing individual is very limited. Consequently, their sources of information from their
surrounding majority come from printed materials
(if they are literate), bilingual deaf signers, and
sign language interpreters. I should also mention
in passing that many hearing individuals have never
met any deaf person, because of the limited contact,
and are not even aware of the existence of such a
minority group in the midst of the global society.
Fortunately, the situation is improving.

Most bilingual deaf signers are postnatal deaf
people, who had already acquired the oral language
(Japanese) when they became deaf, or congenital and
postnatal deaf people who have acquired oralism and
lip-reading at school. Many bilingual deaf signers,
therefore, have learned JSL as a second language
from other deaf people. So, there are deaf people
who are capable of oralism and lip-reading and are
literate but unable to communicate in sign language
with other deaf people, because their parents are
hearing people; consequently, they were sent to an
integrated school or a school for the deaf where
oralism is emphasized and the use of any form of
sign language is prohibited. These people, upon
graduation from such schools, must learn JSL in order
to be accepted by the deaf community in any area,
especially when their parents are not deaf themselves;
otherwise, these deaf people are more often than not
left out from the major activities of the deaf community.

There is also a small group of hearing bilingual signers who are not professional interpreters
nor are they willing to serve as volunteers. They

usually have deaf parents, grew up as native signers, but are too psychologically inhibited to be
of help as interpreters outside the immediate family.
Of course, there are exceptions whereby children
interpret without any hesitation for their deaf parents and hearing husbands or wives interpret will-
ingly for their deaf spouses. In general, however,
the deaf in Japan practice endogamy or marry someone who is hard of hearing; the All Japan Federation for the Deaf makes no attempt to discourage
such practices. This is in sharp contrast with
Singapore, where the Singapore Association for the
Deaf attempts to discourage or dissuade the deaf
from marrying among themselves (13), especially
when the deafness is hereditary.

From the above description of the Japanese
deaf community, it must now follow that there is a
distinct subculture within Japan whose carriers are
the deaf themselves; some of them are not only bilingual but also bicultural; most of them, however,
are confined to their own subculture and never
bother to cross the cultural boundary. Coupled
with this distinct subculture is the fact that
Japan has a history of more than 2,000 years, which
is to say that there must have been deaf people at
least one out of every thousand, statistically, in
every period of Japanese history. Thus, it can be
said that JSL is the independent linguistic product
of the deaf in Japan. One vital difference between
JSL and ASL is that ASL is a relatively recent foreign import (or influence) from France, mixed with
some local elements (14), whereas JSL is a domestic
invention springing from entirely native sources
and following an uninterrupted line of development.
As a native and most natural instrument of expression of thought, then, the varous linguistic forms
of JSL, such as kinship signs reported in Peng (1974),
place name signs, and the characteristics mentioned
earlier, may be expected to mirror the cognitive
universe or worldview of its users — a worldview
which may or may not coincide with that of the surrounding Japanese majority.

In the case of place name signs, for instance,
regardless of whether those explanations from the
deaf are "true" etymologies or secondary rational-
ization, signs whose actual origins have at some
point become obscure, the term "Folk Etymology" as
defined above still holds and succinctly summarizes
the prime relationship between sign language and

culture (see Peng and Clouse 1977). What matters
is how signers themselves explain and interpret the
signs they employ, irrespective of historical verification the study of which falls outside the purview of this presentation. Other relationships between sign language and culture have also been reported. For a detailed discussion of kinship signs,
see Peng (1974).

Further studies regarding JSL and Japanese culture, a realm of investigation that will prove fruitful, should bring into relief the accelerated changes
affecting JSL. These changes are the result of
ever-growing contact between the hearing and deaf
sectors of society, through extralinguistic factors,
be they social, political, economic, or educational.
Let us hope that the changes will be for the better.

Conclusions

First, I believe I have presented enough evidence to conclude that sign language is definitely
a viable alternative to oral language in any human
society. Mechanically, it is suitable to develop
into a written language comparable to any written
language in existence today. (For supporting evidence in this connection, see Peng 1975c; Peng 1976c;
and Peng n.d.a). Since sign language is a viable
alternative to oral language, it must be just as
good a part of language as oral language has ever
been. In our context, then, if oral language, in
line with Malinowski, is the general norm of human
speech activities, sign language, too, is the general norm of human sign activities. Neither case,
however, is a self-contained entity, because each
is intimately related to other cultural systems and
may also be tied to the other as well. It is through
such individual activities (speech or sign) in a
group for the establishment of a norm that changes
of oral language and sign language are conceivable
and made possible (cf. Peng 1976b).

Second, sign language is in many significant
ways a much more complex system in terms of the
techniques for analysis available today, because of
its characteristics of simultaneity, reversibility,
and directionality. But these characteristics are
great assets to communication since subtle meanings
and opposite meanings can be formed easily in sign
language by virtue of those characteristics, whereas
the same effects will have to be achieved with great

effort and difficulty in oral language.

Third, since sign language is a part of language, it is very much a part of culture by any definition; it is a pity that anthropologists have overlooked this important sector of human behavior for so long. It is time that we take sign language seriously and examine the culture of its users without delay.

Fourth, language (oral or manual) is a vehicle of expressions (intellectual and/or emotional) and, therefore, it is a cultural artifact, a man-made convention. In this sense, it must not be confused with man's actual intelligence and other cultural achievements. I must also add that sign language, when compared with oral language, has not been given a chance to develop, phylogenetically speaking; therefore, in some respect, such as vocabulary, it is definitely underdeveloped. Thus, it is all the more important that the use and study of sign language be encouraged. With the current interaction between the deaf and hearing people in Japanese society and the intensified, conscious movement to cultivate JSL (and perhaps other sign languages as well) on the part of the deaf, I believe the day will soon come when JSL (or any other sign language, for that matter) will be just as rich in vocabulary as Japanese (or any other oral language).

Fifth, I am convinced that the superiority of oral language has to a large extent been exaggerated in the past. There are, I think, three interrelated factors for this exaggeration. First and foremost, there has been unawareness, until recently, of the fact that the encoding and decoding processes of information are multi-channeled, not at all limited to the oral-aural channel. There is no reason to suppose, as does Bender, that "the human brain centers for language and the auditory and vocalization mechanism developed together in a complex pattern of interaction and adaptation" (1976:20), when the fact is that the complex pattern of interaction and adaptation involves other senses as well, such as the optic, the tactile, and even the olfactory functions, in connection with motor skills. This is because the nervous system, be it central or peripheral, serves to coordinate body activities in response to external environment (i.e., the outside world) and internal environment (i.e., other systems within the body). Thus, before one can generate a sentence

(to borrow a term from the TG paradigm) or even say
a word, pertinent information has already been pro-
cessed that came in not from the VIIIth cranial
nerve (vestibulo-cochlear) but from the IInd cranial
nerve (optic), when one recognized his superior ap-
proaching, and/or the Ist cranial nerve (olfactory),
when one saw an attractive woman coming, wearing
fragrant perfume, so that he can choose the right
sentence or word to utter. Conversely, then, relevant information that is channeled through the
VIIIth nerve or originates in the brain from its
memory may find its output in the brachial plexus
through the cervical enlargement of the spinal cord,
thereby resulting in gesticulation.

The second factor in the exaggeration of the
superiority of oral language may be found in the
examples of <u>Polly wants a cracker</u> and <u>Einstein invented the theory of relativity</u>; that is, too often
we have confused mental activity (the kind of cognitive processes that can be done without recourse
to language, e.g., painting and music) and language
(oral langauge in particular) to the extent that
language is believed to dominate every human activity. A case in point of this confusion is the
well-known Whorf-Sapir Hypothesis which says that
the language one speaks shapes the world in which
he lives. The truth of the matter, however, is that
langauge (oral and manual) is only one of many cultural systems that man has developed over a long
period of time and that, as such, it has become a
convenient tool — convenient, depending upon how
sophisticated the proficiency of its user is and
how well developed the tool has been — to express
or convey whatever the activity the nervous system
can coordinate from within or without, using all
mechanisms anatomically available. Human cognitive
superiority must be attributed to a general faculty
(the collective fuction of all neurons) that man
possesses — a faculty which is responsible for the
creation of culture and language (oral and manual)
— and not to a particular system or tool itself.

The attitude of ignorance and conceit pointed
out earlier may be the third factor in the exaggerated view of spoken language. More often than
not, we equate "don't know" with "doesn't exist,"
when we compare ourselves with lower primates or
when we compare what we are accustomed to with something that is strange or unfamiliar. Thus, when
Bender compares oral language with sign language,

he gets "the fantastic efficiency achieved by the
coding mechanisms used in speech," for the former,
but "Two obvious speech surrogates — shorthand
transcription and reading of printed text — are
more efficient than sign language" (1976:20), for
the latter, without realizing that many social variables are involved in the surface value which he
took for granted.

Finally, it is hoped that through the pursuit
of rigorous study of sign language, now evident in
many countries, such as the United States, Japan,
Sweden, France, Israel, and others, more insights
can be gained for the true understanding of what
language really is (15).

Notes

1. The problem introduced by Voegelin has since
then become practically moot. The reason, I think,
is partly because, as Voegelin himself puts it, "One
of the results of the growth and prestige of anthropology since World War II is that everybody has
learned what anthropologists mean by 'culture', including pure linguists; anthropological definitions
have been known to be cited for the benefit of anthropologists who are not sufficiently orthodox in
this matter" (1951:364). Also, the issue is closely
tied to the problem of the so-called Whorf-Sapir Hypothesis which in the early fifties excited anthropologists and linguists alike, both pros and cons,
to the extent that a special conference was called
(apparently inspired by Voegelin's suggestion) resulting in the celebrated volume Language in Culture:
Proceedings of a Conference on the Interrelations
of language and Other Aspects of Culture (Hoijer
1954). No serious anthropologist or linguist today
would devote much time arguing enthusiastically one
way or the other on this subject. But the problem
of not reaching an agreement on the term shared by
the practitioners persists as a source of argument
and still lingers today.

2. Linton's definition of culture is "the sum
total of the behaviors of a society's members in so
far as these behaviors are learned and shared"
(1945a:46).

3. An analogy will suffice to explain my contention here. We now know that it took a recent
political venture to introduce Chinese acupuncture

to the United States, even though it has existed
in China for two or three thousand years and has
been known to other parts of the world, like Japan
and Korea, for hundreds of years. But acupuncture,
obviously different from any Western technique to
produce anesthetic effect for operation, cannot be
automatically excluded or ignored from the scope
and aims of anesthesiology or medicine in general,
just because the two techniques are different, even
though they are applied to the same object, the human body. The exclusion would be justifiable, if
and only if one technique were applied to animals
alone and never to humans or it were claimed to be
a cure-all without any empirical evidence.

 4. I have demonstrated (Peng 1975a) what I
mean by true linguistics: "True linguistics, ...,
requires the necessary elements of people, environment, and language. But the TG paradigm lacks all
three, a detriment also reported by other scholars,
e.g., Yngve..." Yngve wrote (1975:1):

> "The dilemma today is characterized by a lack of
> contact between our theories and the results of observ-
> ation; between grammar and the human linguistic activity
> observed in everyday life in our own and other cultures
> ... We see today a crumbling of the transformational-
> generative paradigm. Those linguists who have pinned
> their hopes on it seem disappointed and disillusioned.
> Although much interesting work has been done within
> this framework, it has become more and more evident
> that it suffers from serious internal problems and may
> have outlived its usefulness. Even several years ago,
> in a survey of linguistics from this perspective, edited
> by Dingwall (1971), nearly every author reflects frust-
> ration that the 'standard theory' fails to account for
> many important aspects of natural language, that efforts
> to patch it up seem *ad hoc*, ... One gets the feeling
> that work continues in this framework mainly through
> inertia and for the lack of a more appealing alternative.
> At major linguistic meetings today one senses a feeling
> of crisis in the air."

 5. I have written an extensive review of Chomsky's Aspects of the Theory of Syntax (Peng 1969).
Anyone suspicious of Fraser's claims should read my
review. The book was put together without empirical
evidence; consequently, it became the cause of a
split within the TG paradigm itself.

 6. Those who are interested in this argument

should read Peng (1975b), Bender (1976), Peng (1976a), and Peng (1975c) in that order. I have spelled out clearly what I mean by linguistic innatism, there is no confusion between species-specificity and innateness.

<u>7</u>. Voegelin cites Hull's statement at a seminar at Yale University, <u>circa</u> 1935. It is of interest to note that the idea of teaching such a sign language to nonhuman primates was suggested in the 1930s more or less as a joke, with a tone of caricature. No primatologist, beginning with Allen and Beatrice Gardner (1971) who worked with the chimpanzee Washoe and who successfully taught nonhuman primates to sign, would now take Hull's recommendation seriously. For ASL is not the simplified language intended by Hull; signing chimpanzees and gorillas have learned a system that is much more sophisticated and complex than both Hull and Voegelin anticipated. As a matter of fact, sign language, be it JSL or ASL, is not at all like Voegelin's description and shows characteristics that are far more intriguing than any oral language that has ever been studied.

<u>8</u>. Speculations about the origin of language are not new; they have been debated since ancient Greek times 2,500 years ago, and probably before that as well. Recently (Harnad, Steklis, and Lancaster 1975), the search for the origins of language has been formally declared reopened.

<u>9</u>. For my criticism of this notion of simultaneity, see Peng (1978a).

<u>10</u>. While Stokoe et al. use three types of symbols for one sign (though a segment in my terminology (cf. Peng 1976c), [d] is one symbol which could be converted notationally into three, such as <u>tip</u> <u>alveolar-stop</u> and written as such, each being a component or a feature. Conversely, however, <u>dez</u>, <u>tab</u>, and <u>sig</u> could also be converted into one symbol, instead of three, an idea that has been proposed by Peng (1976c). For example, APPLE:ONION have been said to differ only in location. Thus, they are symbolized as $_3X^{w\cdot}_x$: $_uX^{w\cdot}_x$, respectively. The components in each sign are as follows: APPLE (cheek — hook hand — touch and twisting movement): ONION (mid face — hook hand — touch and twisting movement). Since the two signs differ only in the point of articulation (i.e., <u>tab</u>), each can be

converted into a symbol by drawing upon the convention in phonetics such that the three descriptive
components converge into one symbol and that if one
of the components is different, the entire symbol
is changed or else a diacritic is used. In the case
of APPLE:ONION, if we decide to use X (which tacitly includes the three components, cheek, hook hand,
and touch and twisting movement) for APPLE, then Y
(which also includes the three components, mid face,
hook hand, and touch and twisting movement) can be
used for ONION. Whether or not the notation of one
symbol for one sign (actually one segment) is superior to that of three symbols in a culster (TD^S)
is not at issue here; I have merely pointed out one
distinction between Stokoe et al. and Peng (1976c
and n.d.a).

   The parameter of orientation is something else.
Recall that it is the criterion of differentiation
between SIT:NAME or SHORT:TRAIN. But note that the
distinction between SIT:TRAIN or between NAME:SHORT,
though parallel, is not a matter of orientation but,
rather, a matter of movement (i.e., sig). In line
with this observation, then, the descriptive components of these four signs are: SIT (right "H" —
on the back of left "H" — placed crosswise); TRAIN
(right "H" — on the back of left "H" — rub back
and forth); NAME (middle finger of right "H" — on
index finger of left "H" — placed crosswise); SHORT
(middle finger of right "H" — on index finger of
left "H" — rub back and forth). Such being the
case, I do not think the extra parameter is needed;
all we need to do is to recognize the difference in
contact point as part of dez and tab. If this is
done, then, the distinction between SIT:NAME or
SHORT:TRAIN is a matter of dez and tab and not at
all a matter of orientation. Then, the notation
can follow suit in terms of Peng (1976c and n.d.a),
although I am not sure the notation system of Stokoe
et al. can handle the signs without the extra parameter. This is another difference between Stokoe
et al. and Peng (1976c and n.d.a).

   11. The term "sign symbolism," as used in the
present context, should not be taken as pleonastic
nor should it be equated with "iconicity." For one
thing, if sign symbolism is pleonastic, so is sound
symbolism, because sound may also be a symbol. For
another thing, sign symbolism and iconicity are not
the same thing. Sign symbolism has little resemblance to its referent, but iconicity usually bears

some resemblance between a symbol and its referent (or event).

12. For another thing, note that KNOW and NOT KNOW in JSL need not show sign symbolism, just because they do in ASL. This is to say that sign symbolism should not be equated with universality; one must not presume that since KNOW and NOT KNOW pertain to the head at some point in ASL, they must do likewise in JSL; since I have said (with reference to Reversibility) that they are signed on the chest in JSL, I must have forgotten something. No, I didn't forget anything. It is just JSL and ASL are two different languages and sign symbolism in one does not have to correspond to sign symbolism in the other. Note also that KNOW and THINK need not be synonymous in sign language (JSL in particular) any more than they both should show sign symbolism pertaining to the head. In some oral languages, people "think" by referring to the heart.

13. I am a life member of the Singapore Association for the Deaf. I was given a report, several years ago, by the Secretary of the Association, describing three marriages between three sisters and three young men all of whom are deaf. A match-maker first introduced the sisters to the young men who, then, fell in love at first sight. When the Association learned of this arrangement, counseling was provided immediately aiming at discouraging the marriages for fear of hereditary deafness. But apparently love at first sight was too strong. So, the Association blessed each couple and wished them luck. A year or so later, the first sister had a baby; luckly the baby was normal. Then, the second sister's baby was born, also with normal hearing. But the third sister gave birth to a baby found to be hearing-impaired.

14. In a recent report, entitled "Historical Bases of American Sign Language," Woodward concluded that although ASL was originally taught to the American deaf in the mold of French Sign Language, new evidence suggests that there had been a variety of native signs in the United States which were gradually mixed with the imported artifact by the American deaf and that contemporary ASL is a mixture of native and foreign signs. It would be of interest to see if this claim holds beyond the level of individual signs; that is, if partial grammar of the native sign language has been mixed with

that of the imported one to make up contemporary ASL currently used in the United States and Canada. A close analogue in oral language is Japanese which has mixed Chinese and native grammars together to a considerable extent.

15. I am grateful to Gordon W. Hewes of the University of Colorado for material improvement in this chapter.

References Cited

Bender, M. Lionel
    1976 "In Defense of Linguistic Innateness: Reply to Peng" Language Sciences, August, pp. 19-20.

Christopher, Dean A.
    1976 Manual Communication, Baltimore: University Park Press.

de Saussure, F.
    1955 Cours de Linguistique Générale, Paris: Payot.

Eimerl, Sarel and Irven DeVore
    1974 The Primates, Life Nature Library, New York: Time-Life Books, Time Inc.

Fraser, C.
    1966 "Discussion of D. McNeill's 'The Creation of Language by Children'" in J. Lyons and R. J. Wales (eds.), Psycholinguistics Papers, Edinburgh: Edinburgh University Press.

Frishberg, Nancy
    1975 "Arbitrariness and Iconicity: Historical Changes in American Sign Language" Language 51(3).696-719.

Harnad, S., H. D. Steklis, and J. Lancaster (eds.)
    1975 Origins and Evolution of Language and Speech, New York: The New York Academy of Sciences.

Hill, A. A.
    1971 "What is Language?" in Charlton Laird and Robert M. Gorrell (eds.), Readings about Language, New York: Harcourt, Brace, and Jovanovich, Inc.

Hoijer, Harry
    1948 "Linguistic and Cultural Change"
          Language 24(4).335-45.

Hoijer, Harry (ed.)
    1954 Language in Culture: Proceedings of a
          Conference on the Interrelations of
          Language and Other Aspects of Culture,
          AA, Vol. 56, No. 6, Part 2, Memoir No.
          79, December, Washington D. C.: The
          American Anthropological Association.

Kluckhohn, Clyde and William Kelly
    1945 "The Concept of Culture" in Ralph Linton
          (ed.), The Science of Man in the World
          Crisis, pp. 78-106, New York: Columbia
          University Press.

Lehmann, Winfred P.
    1973 Historical Linguistics: An Introduction,
          New York: Holt, Rinehart, and Winston,
          Inc.

Linton, Ralph
    1945a The Cultural Background of Personality,
          New York: Appleton-Century.
    1945b "The Scope and Aims of Anthropology" in
          Ralph Linton (ed.), The Science of Man
          in the World Crisis, pp. 3-18, New York:
          Columbia University Press.

Opler, Morris E.
    1949 "Words without Meaning or Culture" Word
          5.42-4.

Peng, Fred C. C.
    1969 "Review of Aspects of the Theory of Syn-
          tax by Chomsky" Linguistics 49.92-128.
    1974 "Kinship Signs in Japanese Sign Language"
          Sign Language Studies 5.31-47.
    1975a "Sociolinguistics Today: In Lieu of An
          Introduction" in Fred C. C. Peng (ed.),
          Language in Japanese Society: Current
          Issues in Sociolinguistics, pp. 3-31,
          Tokyo: The University of Tokyo Press.
    1975b "On the Fallacy of Language Innatism"
          Language Sciences, October, pp. 13-6.
    1975c "The Deaf and Their Acquisition of the
          Various Systems of Communication: Specu-
          lation against Innatism" in James Macris
          and Walburga von Raffler-Engel (eds.),

Child Language 1975, pp. 225-46, Word
Vol. 27, Nos. 1, 2, & 3.

1976a "Linguistic Innateness: Reply to Bender"
Language Sciences, October, 34-5.

1976b "A New Explanation of Language Change:
The Sociolinguistic Approach" Forum Lin-
guisticum 1(1).67-94.

1976c "Sign Language and Its Notational System"
in Peter A. Reich (ed.), The Second
LACUS FORUM 1975, Columbia, S. C.: Hornbeam Press, Inc.

n.d.a "More on Sign Language and Its Notational
System" in Fred C. C. Peng (ed.), Toward
a Science of Body Movement, Hiroshima:
Bunka Hyoron Publishing Company, forth-
coming. Also in Fred C. C. Peng (ed.),
Aspects of Sign Language (in Japanese),
1978, Hiroshima: Bunka Hyoron Publishing
Company.

n.d.b "Urbanization and Language Sciences: The
Japanese Case" paper presented at the
AAAS Bicentennial Meeting in Boston,
1976. Also in Fred C. C. Peng (ed.),
Language and Context, 1978, Hiroshima:
Bunka Hyoron Publishing Company.

n.d.c "Introduction" in Fred C. C. Peng (ed.),
Toward a Science of Body Movement, Hiroshima: Bunka Hyoron Publishing Company,
forthcoming.

Peng, Fred C. C. and Debbie Clouse
1977 "Place Names in Japanese Sign Language"
in Robert J. Di Pietro and Edward L.
Blansitt, Jr. (eds.), The Third LACUS
FORUM 1976, Columbia, S. C.: Hornbeam
Press, Inc.

Peng, Fred C. C. and Peter Geiser
1977 The Ainu: The Past in the Present, Hiroshima: Bunka Hyoron Publishing Company.

Riekehof, Lottie
1963 Talk to the Deaf, Springfield, Mo.:
Gospel Publishing House.

Stokoe, William C., Dorothy Casterline, and Carl
Croneberg
1965 A Dictionary of American Sign Language
on Linguistic Principles, Washington
D. C.: Gallaudet College Press.

Supalla, T.
      n.d.   "Systems for Modulating Nouns and Verbs
             in ASL" paper presented at the Confer-
             ence on Sign Language and Neurolinguis-
             tics, September 24-26, 1976, Rochester.

Tylor, E. G.
      1874   Primitive Culture, London: John Murray.

Voegelin, C. F.
      1949a "Linguistics without Meaning and Culture
             without Words" Word 5.36-42.
      1949b "Relative Structurability" Word 5.44-5.
      1951   "Culture, Language, and the Human Or-
             ganism" Southwestern Journal of Anthro-
             pology 7.357-73.

Voegelin, C. F. and Z. S. Harris
      1945   "Linguistics in Ethnology" Southwestern
             Journal of Anthropology 1.455-65.

Woodward, J.
      n.d.   "Historical Bases of American Sign Lan-
             guage" paper presented at the Conference
             on Sign Language Neurolinguistics,
             September 24-26, 1976, Rochester.

Yngve, Victor
      1975   "The Dilemma of Contemporary Linguistics"
             in A. Makkai and V. B. Makkai (eds.),
             The First LACUS FORUM 1974, Columbia,
             S. C.: Hornbeam Press, Inc.

Code and Culture

Nancy Frishberg

Linguists, that is people engaged in the study
of Language, have traditionally been confused with
polyglots, who study languages. Sign language linguists are engaged in the study of Language which
is produced and perceived in the manual-visual modality. Although many people have heard about American Sign Language (ASL, or sometimes Ameslan) because of its adoption in experiments with non-human
primates (see Fouts', Miles', and Patterson's articles, this volume), there remain some misunderstand-
ings about the nature of sign language.

While the distinction between a manual-visual
language and gestures — non-verbal, non-vocal cues
accompanying speech — might be recognized and ac-
knowledged, the impression persists that sign language is "the same everywhere," a universally understood and uniformly articulated code. In fact,
indignation is the usual reaction to the linguist
who attempts to counter this myth of sign language
universality.

The purposes of the present paper are therefore twofold. The first is to demonstrate that
despite the basic similarity in modality of production and perception, naturally occurring sign
languages of the world are necessarily distinct
from one another and vary in their formational
structures in ways that are analogous to the variation in spoken language sound structures; that is,
sign languages differ from one another in arbitrary
and probably irreconcilable ways. Secondly, despite the gross differences in formational characteristics, the sign language and spoken language of

a given culture can be expected to share some relationships. There is a generalized influence of
culture reflected in both languages; or, to put it
another way, some of the culturally relevant categories and classifications become apparent in both
languages in similar ways.

Examples in this work will be drawn primarily
from American Sign Language, but occasionally will
come from Chinese Sign Language and from the indigenous sign languages of Rennell Island, Providence
Island, and Adamorobé, Ghana (1). The arguments
here built on languages from geographically and
culturally distinct areas will explicate some of
the many ways that sign languages might differ from
one another.

Language Universals and Phonology

As has been mentioned previously by other researchers and myself, American Sign Language (ASL)
is descended from a lineage quite separate from the
oral language tradition of English. ASL is known
to be descended from an older form of French Sign
Language, although more recent evidence suggests
that modern ASL may have a large indigenous American component and is not solely comprised of signs
derived from the French Sign Language which was itself standardized and instituted as the official
sign system of the first public school for the deaf
(Lane 1976; Woodward 1976). Throughout the his-
tories of various sign languages we find the factor of the educational systems adopted in a particular region influencing the character of the sign
language. Today French Sign Language and American
Sign Language differ widely from one another, as do
the other languages also descended from the Older
French Sign Language (Russian, Irish, etc.). Furthermore, British Sign Language constitutes a completely separate signing lineage even though the
written and spoken languages of Britain and the
United States are essentially the same, that is,
English. These historical facts alone might serve
to show the non-universality of sign language.
However, there are many respects in which the sign
languages are similar, which are hidden by the
statements above.

Characteristics of the various sign languages
can be found that are shared despite the differ-

ences in origin or evolution. The articulation
area for each sign language investigated so far appears quite uniform. The top of the head seems to
mark the upper edge. The lower boundary, somewhat
less rigorously applied, is the waist. The arms
extend sideways and frontwards, not quite full.
The space for a single signer may be somewhat distorted on interaction with another signer who is
nearby (Lacy 1974). (The full description of this
distortion has not been written yet.) Notice, however, that the similarity of the signing space does
not constitute linguistic universality — the exact
combinations of words and elements within words are
still distinct.

   The signs can be considered to be a composite
or simultaneous realization of at least four distinct parameters first given by Stokoe (1960),
where he mentioned three and assumed the fourth. A
sign is composed of a specific handshape (designa-
tor = dez) held in a particular location (tabula =
tab) executing some movement (signation = sig).
The fourth parameter specifies the orientation of
the signing hand. The breakdown of signs into at
least these four parameters seems to be quite consistent across different sign languages.

   Sometimes signs use both hands. Each of the
sign languages we have seen has signs of both the
one-handed and two-handed variety. Usually, the
right hand will execute the movement of a one-handed
sign. Also, for those signs that are not symmetrical among the two-handed signs, the right hand will
take the active role. The majority of signers investigated are right-handed (2). Previous researchers have shown that ASL obeys two constraints
with respect to two-handed signs (Battison 1974).
First, when the two hands are both moving, they
must be identical in configuration and location.
They must carry out the same movement (e.g., WHICH,
EXCITED, GERMAN) (3). Second, when the two hands
have different configurations they must be in the
relation of active hand - base hand. That is, one
hand moves but the other must not. It can serve as
the base hand for the active hand though. Usually,
the dominant hand (right hand for most people) will
take the active role (e.g., HOUR, DUTY, PRACTICE,
TURTILE). These two principles are called the Symmetry Constraint and the Dominance Constraint, respectively.

With the four parameters it is possible to distinguish signs from one another. Minimal pairs can be found in ASL based only on differences in one of the four parameters. Several of these will be given below. Furthermore, the structural properties outlined above seem to hold true for most sign languages investigated thus far, although a really careful comparative study needs to be done.

What is the substance of a claim that sign language is universal? We can understand it in several ways. On one reading it would mean that there is a consistent way of producing signs that will be consistently comprehended; that is, that signs are uniform everywhere. We would then expect that the same handshapes, locations, movements, etc.,would appear for a particular meaning everywhere. Another way of understanding the claim is that there are predictable relationships between, for example, differing handshapes among different groups (for example, geographically distinct areas). We can show that neither version of the claim of universality fits the evidence from sign language of the deaf.

 If we take only one parameter, handshape, we can demonstrate both of these points. American Sign Language has about 40-45 different handshapes that are used at different times. These are the "phonetic" shapes. Of these only about 19 are distinctive - these are the "phonemic" shapes. These 45 shapes, however, do not exhaust the possible shapes that a human hand can form.

The sign languages of China and Japan both use at least one shape which does not exist in ASL; namely, ring finger extended from the fist (Figure 1). This observation makes no prediction about the relative frequency or difficulty of one shape over another. It merely suggests that there are configurational elements smaller than the whole word in sign languages and that these elements are systematically different from language to language.

The Chinese Sign Language (CSL) handshape (which is used in signs like CSL RIVER, OLDER-BROTHER, etc.) — middle finger extended from the fist — is likewise unacceptable in ASL. The handshape which is also taboo among hearing people in our culture does not enter into the formation of individual signs in ASL, but in Chinese culture apparently the same restrictions do not occur since this element enters productively into the sign vocabulary. So, here we have an example of a structural property of ASL which reflects the more general gestural community's conventions. But the convention does not hold necessarily for other gestural or signing communities. The CSL signs RIVER and OLDER-BROTHER never could be pronunciations in ASL. There are shapes, locations, or movements that are not taboo for Americans, hearing or deaf, but which would violate gestural taboos for persons from other cultures.

Bellugi and Klima (1975) have pointed out that the distinction between U and V, index and middle fingers extended (non-spread versus spread), is one which can separate and identify different words (UNIVERSITY vs. 2-WEEKS-FROM-NOW) in ASL, but this same distinction in Chinese Sign Language does not exist (compare Figures 3 and 2). It cannot be used to show the difference in meaning between two Chinese signs. Similarly, the shapes V and 5 (two fingers, index and middle, extended from fist, as in Figure 2, versus open spread hand, as in Figure 4) appear to be related to one another, i.e., substitutable for one another in some contexts, in the indigenous sign language of Adamorobé, Ghana (4). The sign for "Friday" in Adamorobé Sign Language (AdaSL) can be made with two fingers contacting the forehead either while the others are clenched in the fist or while all the other fingers are open and relaxed. The same optional alternation is not possible in ASL. In fact, V and 5 can distinguish signs of quite different meaning (TWO-WEEKS vs. NICE/CLEAN).

We predict then that sign languages together have a set of potential handshapes from which any one language chooses a subset. The same prediction will also work for the other three parameters. We can show, for example, that some signs in one language violate formational principles in another. The better we understand the structure of each language, the more specific we can be about how the signs of one language violate the structure of another.

Providence Island Sign Language (PISL), an indigenous sign language of a Caribbean community,

Figure 1:   Chinese Sign Language
            Handshape

Figure 2:   ASL shape V   Figure 3:   ASL shape U

Figure 4:   ASL shape 5

violates the conventions for ASL, for example.
There are signs in PISL which are distinguished only by the shape and movements of the mouth (e.g.,
PISL DOG vs. PIG). American Sign Language tends to
avoid using mouth movement distinctively and in
fact historically seems to change to eliminate non-
manual articulatory mechanisms in signs (Frishberg
1975). So, the distinctive use of mouth movements
in PISL is a violation of the structure of another
sign language, namely, ASL.

Two more examples show how other movement vio-
lations can distinguish sign language systems from
one another. Adamorobé Sign Language uses a locat-
ion-movement combination in CHASE which does not
exist in ASL, namely the elbow surface repeatedly
contacting the side of the body. The elbow and
lower arm areas are acceptable as <u>locations</u> in ASL
(as in TEMPT, CRACKER, COUNTRY, PO<u>O</u>R, etc.) but the
elbow is never the active component. AdaSL however
apparently can use the elbow as the active articulator and in this respect is structurally distinct
from ASL. Chinese Sign Language like ASL allows
change in handshape from one configuration to another to constitute the movement parameter of a
sign. In both languages only some shapes may be
the initial or final shape of a sign. However, CSL
uses the change from thumb-extended to pinky-exten-
ded as one of those movements (e.g., CSL SUSPECT).
These two shapes are alone each acceptable in ASL
but never alternate with each other as the movement parameter of a sign.

The above examples are intended to support the
argument that sign languages of the deaf are not
universal. In fact, although they share many properties, including the generalized notion of signing
space and perhaps the constraints on dominance and
symmetry, they are composed of different word-
building component parts (specifications for handshape or movements, etc.) and thus the sub-lexical
structures of different sign languages are different. One more kind of evidence that sign languages
are not universal and indeed differ in specific
ways comes from the discovery of cross-linguistic
homonyms. Occasionally the same sign will appear
in different languages and will have different
meanings. That means that the configuration of
sub-lexical parts (handshape, movement, location,
etc.) is the same but the combination produces a

Figure 5:   The number 3 in American Sign
            Language

Figure 6:   The number 3 in American gestures, Ren-
            nellese Sign Language, Adamorobé Sign
            Language

Figure 7:   The number 3 in some Chinese Sign
            Language forms

different meaning. For example, ASL SECRET is
FATHER in Chinese Sign Language. CSL CANTON is the
same form as AdaSL WAIST-BEADS. The Russian Sign
Language form MOSCOW is the same as ASL MENSTRUAT-
ION. Kagobai's sign for SIBLING looks to ASL signers like it represents TWO-PEOPLE-SEPARATE. This
sign is discussed in greater detail later. So,
what we saw previously is that there are structural
incompatibilities between different sign languages
which prevent them from being universally understood. Here, we find that even when the structures
are compatible, so that the same formation is allowed in different language communities, the meanings have already been arbitrarily assigned and are
different.

Language Structure and
Sign Language Semantics

We have just seen that sign languages form
cross-linguistic "homonyms" in unpredictable ways.
What if sign language were universal? We would expect a particular meaning to be represented in the
same way whether made in America or China. We might
predict that some culturally neutral terms like
those for numerals would be identical from language
to language. Let us look at several numerals: The
ASL sign for both the ordinal and cardinal number,
"three," is made by showing the extended thumb,
index, and middle fingers (cf. Figure 5). The sign
languages of Kagobai of Rennell Island, of the vil-
lage of Adamorobé, Ghana, and the gesture of hearing people in the United States form "three" with
the extension of index, middle, and ring fingers
(Figure 6 shows thumb holding pinky against palm).
Chinese Sign Language uses both this form for simple
counting but also occasionally uses the middle,
ring, and pinky fingers extended handshape (Figure
7, thumb and index contacting in a circle) in terms
like CSL NAME, WEDNESDAY (3rd day), etc. Notice
that each of these handshapes shows "three" by holding up three contiguous fingers. We cannot predict
from this the universal manual representation of
the notion "three," but we probably would be correct in predicting that any sign language which has
a distinct numeral for "three" will use a form with
three contiguous fingers distinguished. This last
statement is actually a very cautious attempt at a
typological universal (as was presented in Greenberg

1966) and says nothing about the exact formation of
the numeral 3 in any particular sign language. It
in fact refrains from predicting that the three
contiguous fingers will be <u>extended</u> from the fist,
and allows the possibility <u>that the</u> fingers will
instead be bent down to the palm from an extended
position. It does, however, predict that no sign
language will use, for example, index, ring, and
pinky extended from a fist to mean "three."

Moreover, we should recognize that the shape
(shown in Figure 6), which in Rennell and Adamorobé
and English gesturing means "three," in ASL means
"six" (5). The identical handshape shown to
signers of different languages will communicate
different numbers. Similarly the Chinese Sign Language form shown in Figure 7 is interpreted by ASL
users as the number 9 (6).   So much for our universal sign language; we can't even count to ten
without ambiguity. Signers from distinct tradi-
tions have distinct representations for the numerals.

To carry this point one step further, ASL uses one hand to show all numbers. In the same way
that fingerspelling in ASL is a one-handed process,
so citation of numbers as well. However, Adamorobé
Sign Language uses one hand for numbers from one to
five, and two hands for six through ten. Optionally SIX and SEVEN can occur without the neutral
shaped base hand. Compare the shapes in Figures 8
and 9. One handed variants of AdaSL forms 6 and 7.

The AdaSL forms for "six" and "seven" are interesting from another point of view. AdaSL SIX
is formed by contacting palm surfaces of a flat up-
raised hand with the pinky extended from the fist
shape. SEVEN extends the pinky and ring fingers
and contacts these two on the dominant hand's
palm surface. Notice the analytic character of
the two forms: Active hand represents "five" and
base hand "one" or "two," repsectively. Note also, however, that the "five" in AdaSL SIX and
SEVEN is distinct from AdaSL FIVE when signed in
isolation. The combining form is an open hand a-
gainst which another shape can contact (e.g., AdaSL
SIX through NINE). The citation form AdaSL FIVE is
made by contacting all the fingers to the thumb.
Moreover, the active handshape which can be analyzed as meaning "two" in AdaSL SEVEN is distinct

from AdaSL TWO and is a shape which does not occur
in any ASL form (7).   What we have found is a way
in which ASL and A̅damorobé sign specifications are
incompatible.   Even then in these cases where the
semantic value of the sign is a universally understood and expressible notion, the shapes of the
surface forms are quite distinct.   We have examined
a few numerals in different sign languages and seen
that each of the languages has a distinct way to
express the notion common to all.   The similarity
among them is the extension of three fingers, but
the specification of <u>which</u> three is not predictable.   We see that the sign language and oral language of a given locale do not necessarily share
the manual representation of a particular numeral.
And what's more the numbers are relatively arbitrary within a given sign language.   The ASL form
for "six" does not use six extended fingers.   When
considered within the pattern of ASL numerals, SIX
is relatively motivated - it fits the pattern of
numbers (compare ASL SIX, SEVEN, EIGHT, NINE in
Figure 8).   But there is not an apparent or necessary connection between the meaning "six" and the
ASL sign SIX, so we can say that this sign is an
arbitrary form.

　　To take another form, consider the signs for
"sibling."   Each of the sign languages investigated
thus far has at least one form for this term.   If
we first describe the signs it will be clear how
many possible representations are possible, and,
what's more, how those representations cue different information in the different cultures.   ASL
touches the index finger to the forehead and then
brings the two index fingers side-by-side pointing
outward for the sign BROTHER.   The sign SISTER
makes the first contact on the cheek rather than
the forehead.   This compound sign has two components, the first designating the gender of the sibling and the second indicating SAME.   The form of
the sign mentioned here is the assimilated form.
Compare the AdaSL form for SIBLING which simply
shakes the hand back and forth with index and middle fingers extended, the hand held in front of the
body, palm toward signer.   Here, too, we find
AdaSL's SAME utilized in the formation of this kinship term.   However, now consider Kagobai's term
OPPOSITE-SEX-SIBLING.   His sign uses the two hands
with index fingers pointing upwards, back sides together.   The hands separate so that one is close to

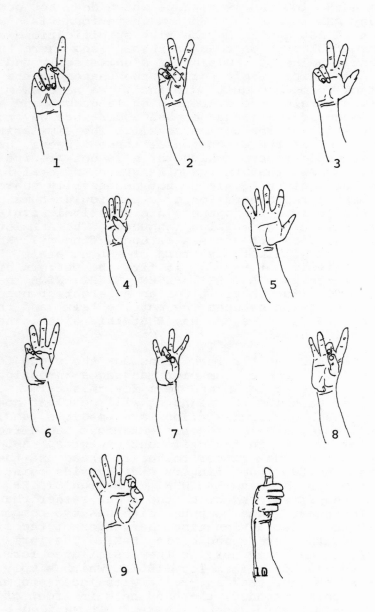

Figure 8:   American Sign Language numbers 1-10

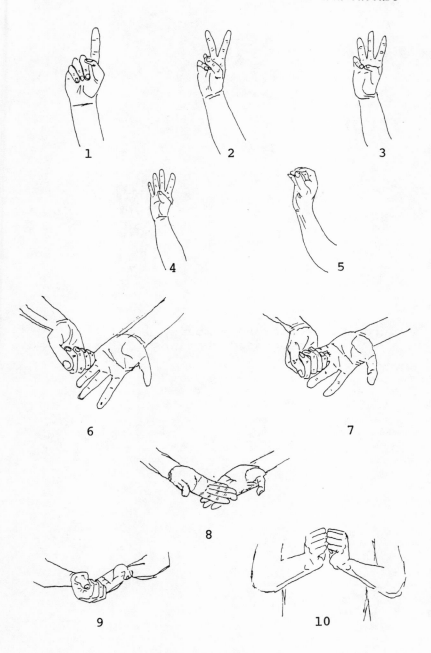

Figure 9:   Adamorobé: Sign Language numbers 1-10

Figure 10:   Adamorobé numbers, 6 and 7
              One handed variants

the body and the other further away, palm surfaces
still face in opposite directions. This sign is
used for a sibling of the opposite sex and is representative of the Rennellese cultural tradition of
brother-sister avoidance. The crucial and distinguishing characteristic of the brother-sister relationship for Kagobai, the only deaf individual on
Rennell, is the avoidance taboo. Chinese Sign Language uses different handshapes for the male and
female kin terms, but in general contacts the mouth
with the appropriate shape. But the terms MALE-
SIBLING and FEMALE-SIBLING are not possible alone;
rather they must be accompanied by an indicator of
the age relationship between the siblings. The CSL
forms then are compounds of two elements: first either MALE-SIBLING or FEMALE-SIBLING, then OLDER-
THAN or YOUNGER-THAN. The cultural tradition of
the larger community (hearing as well as deaf) certainly plays a role here. The signs for the kinship terms described above are quite different in
each language both in their exact formation ("phonetic form") and also in their morphology. The an-
alyzable components of the compound signs are not
simply relexification of the same elements in different languages. That is, the ASL compound MALE +
SAME or FEMALE + SAME, understood by signers in the
United States as "brother" and "sister," would not
be appropriate for communicating to Kagobai of Rennell Island, whose notion of "opposite sex sibling"
necessarily includes mention of the avoidance taboo. Nor are the ASL form's meaning elements possible for Chinese signers who not only require the
gender of the sibling to be specified but also the
age relation as well.

We have begun here to mention several ways
that the cultural tradition of the larger community
is relevant for understanding the sign language's
analytic morphology. That is, the claim here is
not simply that the sign languages differ from one
another in arbitrary and unresolvable ways, but also that the sign language tradition of a particular
locale is related to the cultural tradition in important ways that likewise prevent the signs from
being universally intelligible. The arguments made
here talk about the structural and semantic-cultur-
al incompatibilities in the currently extant sign
languages.

The next section shows that this relationship

between the spoken language, the sign language, and
the shared culture is much more complex than has
been previously thought.

Metaphor: Cultural or Linguistic

One question which often arises in an initial
discussion of sign languages is the issue of the
expressive power of a sign language. For no obvious reason, some people naively assume that a language which uses the manual-visual modality will be
tied to more concrete or specific notions and will
lack lexical items referring to ideas or generalizations or abstractions. To refute this myth we
can simply mention that ASL has a fully developed
vocabulary referring to emotions and feeling
states (e.g., JEALOUS, HUNGRY, DROWSY, etc.), un-
real states (DREAM, FANTASY, etc.), past and future time (more on this later), and moral or meta-
linguistic states (JUSTICE, PUNISH, CHEAT, IRONY/
SATIRE, WORD, SENTENCE, etc.).

What is even more interesting to us on this
occasion is the use of metaphor and extension of
meaning. Previous researchers have noted that
without metaphor every distinct idea and action
would require a separate lexical item to name it
(Ullman 1966). The extension of one word's meaning to refer to other ideas and objects is an important organizing principle in language structure
and seems to be a universal of all human languages.
If we look at the relationship between a spoken
language and the sign language of the same geographic environment, we find that there are some important relationships which have been overlooked or
ignored in previous work. Specifically, we can
show that the cultural relationships are shared observations between the hearing and deaf communities. That is, the deaf population uses many of
the same metaphors in the sign language that the
hearing population uses in the spoken language. We
can look at three distinct types of lexical exten-
sions and show correspondences between the oral and
signed languages. In each of these cases, however,
the argument can be maintained that the sign language has borrowed the metaphorical extension from
the spoken language. A fourth type of example will
however make the distinction between oral language
borrowing and shared culture influence even more
explicit.

Consider the English word <u>pass</u>. When used as a verb, it refers to the motion of a self-moving object in relation to another object or location. This might be called the literal meaning of the word, as in

(1) Our car passed yours on the freeway.

The metaphoric extension of this sense of <u>pass</u> is seen in

(2) I passed the math test.

Now, let's compare the ASL sign PASS. This sign is made by holding the two fists with thumbs up and moving the dominant hand from nearer the signer's body to further away, extending the arm and passing close to the static base hand. This sign is related to a number of other signs all of which have meanings based around the actions and relations of movable objects, e.g., FAR, CHALLENGE, CHASE, FOLLOW, AVOID, EVADE, etc. So, the formation of the sign PASS has no physical relation to the sound of the English word <u>pass</u>; these two formatives are part of separate "phonological" and morphological systems. But the sign PASS with a minor modification in movement called "sharp" — tensing of the arm muscles with rapid onset of action — can be used for the metaphorically extended meaning of "pass a test," "pass a course," to receive a "passing" mark. It is not clear whether the meaning here has arisen spontaneously or if it was borrowed from the English usage. One suspects borrowing, since this semantic realm, i.e., school terminology, is one in which deaf-hearing interaction occurs frequently. Note that other metaphoric uses of English <u>pass</u> do not necessarily translate to a sign which is identical to or immediately related to the sign here glossed PASS. For example, the phrase "to pass out" meaning "to lose consciousness" is translated into ASL as FAINT.

The same mechanism that relates literal PASS to metaphoric PASS relates a number of other literal and metaphoric pairs: KNEEL, WORSHIP (= "adore"); AGREE, BECOMING (where this sign means specifically that a garment or physical attribute "agrees" with one's appearance), etc.

Note that in Adamorobé Sign Language, which is surrounded by the Twi-speaking-hearing culture, the sign EAT is also used for the sense "use" or

"spend" as in the Twi verb d<u>i</u> 'eat'. Also, the ri-
tualized insults used by Adamorobé signers may be
borrowed in part or whole from their Twi-speaking
relatives and neighbors. One such insult asserts
that the target person has "tiny eyes." I am informed that Twi speakers also consider "tiny eyes"
to be an undesirable characteristic. The Ghanaian
examples, then, illustrate again the shared forms
of linguistic expression between spoken and signed
languages of a single community. Notice, here, especially that the Adamorobé signers have not traditionally been educated to speak, read, or write Twi
(or English, for that matter) and whatever influence exists from Twi on the sign language is introduced into the linguistic situation by hearing
signers who are also speakers of Twi. Examples
like those above are probably a result of this hearing Twi-speakers' influence on the sign language.

A second kind of influence the shared culture
of the dominant spoken language community can have
on the signing community is related to the first,
illustrated above. The first type mentioned lexical items which seem to exist in both spoken and
signed language in literal and metaphoric uses. The
second type of influence of spoken language culture
on signers can be seen in the morphology of the
sign language. In ASL, groups of signs which share
consistent meaning components and formational elements are often said to be morphologically related.
So, for example, ASL locates the majority of lexical
items for verbalization and communication in the
area near the mouth. (See Sign Symbolism illustrated
by Peng in this volume.) Thus, signs for SPEAK,
TELL, TALK-CONTINUOUSLY, TRUE, FALSE, LIE, ORDER,
ANSWER, CONVERSE, DIALOGUE (= 'two people talk'),
ANNOUNCE, TATTLE-TALE, etc. are all made in the
area of the mouth. A smaller proportion of signs
for SIGN-LANGUAGE, WORD, ASK/QUESTION, LECTURE,
GOSSIP, and SENTENCE seem to be exempt from this
oral culture domination. The claim here is not
that English language has had an influence on ASL
but, rather, that spoken language culture assumes
the central role of the oral track in realizing linguistic utterances, and that ASL has adopted that
metaphor to refer to utterances made by the hands
as well as those made by the mouth (<u>8</u>).

Another example of this morphological reflex
of the general cultural influence on sign language

might be found in the ASL signs made on the nose.
The sign SMELL ('smell' or 'odor') is made in the
area of the nose. The sign STINK 'bad odor', likewise, uses the nose as the location and, furthermore, has the extended meaning comparable to the
English word ('to be in bad repute'). A number of
other signs also use the nose as their location and
many of these express the unfavorable connotations
noted for STINK (9).   For example, JUNK, SCANTY,
LOUSY, SNOBBISH, $\overline{\text{I}}$GNORE, DON'T-CARE, ODD/STRANGE,
BORED, UGLY, etc. We might speculate on the source
for this collection or "family" of signs by noticing that English speakers use a gesture like the
sign STINK to show their distaste. The prolifera-
tion of signs of distinctive movements and handshapes in the target location, however, cannot be
motivated by any connection with English. The
claim here, once again, is that the culturally relevant value of the word <u>stink</u> in English and the
gesture of holding the n$\overline{\text{os}}$e are both reflected in
the ASL lexical item STINK, and are likewise pres-
ent in the other signs made in the same location.
There is no evidence from ASL that the sign STINK
was the basic sign from which the others have been
derived; this example should not be taken as supporting a theory of gestural origin of all signs.
Rather, the culturally present metaphor of dis-
tasteful quality is expressed in the English word
and gesture and in the whole group of signs made
in the location of the nose, including the sign
which in form is identical to the English gesture,
and whose gloss is given here by the English word
of similar meaning.

   In some cases, the meaning which a lexical item
carries in ASL is much like one would expect given
the English gloss. The example of STINK above is a
word in this category — the literal and metaphoric
meanings of the English word are appropriate for
the ASL item STINK as well. There are other cases,
however, where ASL has borrowed a lexical item from
English and modified the meaning slightly. One
case which is particularly clear is the compound
sign BRAIN-WASH. The fact that this sign is a loan
translation from English is evident from the gloss
and the usage among college-educated signers. The
initial element in the compound is a form of the
sign THINK and the second element is an unassimila-
ted form of the sign WASH, $\overline{\text{A}}_{\text{a}}$ A$_{\text{b}}$ $\overset{\text{o}}{\text{x}}$. The meaning
change in the sign BRAIN-WASH removes the sense of

external coercion which the English word has. So
English allows a sentence like:

(3) Sgt. White was brainwashed by the enemy.

or even:

(4) Sgt. White is brainwashed about basketball.

where the implication might be that she can't stand
seeing the hoops or whatever. The meaning of
BRAIN-WASH in ASL is something more like "obsessed
with" or "very interested in." So we would find:

(5) SGT. WHITE BRAIN-WASH FOR BASKETBALL

where the implication is that basketball is her fa-
vorite game; she prefers it to all others, or something equally positive.

Occasionally, ASL users will borrow a whole id-
iomatic phrase from English. These borrowings may
be marked in one of several ways. The whole phrase
can be preceded by scare quotes using the sign
QUOTES (see below for a more complete discussion of
this sign), or one or more words in the idiom can
be translated into signs using lexical items that
are particularly marked as a hearing person's term.
So, for example, one man was describing a three-
year-old deaf child's proficiency in acquiring sign
language and said:

(6) THREE YEAR OLD AND (HE) SIGN ALIKE *CRASY
                                        $3G_{<}^{⌀}$
'(Only) three years old and he signs like
crazy'.

This example illustrates several points. First,
the signer has used a hearing person's gesture for
"crazy," the index finger pointing toward the temple and making a circling movement. In this way he
has marked the phrase "like crazy" as being the
English phrase. Consider alternatively the meaning
of the sentence if the signer had used one of the
standard ASL signs for CRAZY $3C^{ω}$:

(6')...AND (HE) SIGN ALIKE CRAZY

'and he signs like $\left\{ \begin{array}{l} \text{(he's) crazy} \\ \text{a crazy (person)} \end{array} \right\}$'

In the sentence, where we have replaced the English gesture by a native ASL sign, the meaning is no longer a flattering comment on the skill of the three-year-old signer. Thus, the use of an obviously borrowed gesture accomplishes a double task — it marks the phrase as one borrowed from the hearing world, and thus the meaning 'crazy' is to be interpreted as an English idiom.

Languages without a written tradition are probably more frequent statistically among the world's languages than those with a written language tradition. (Having no native written tradition says nothing, of course, about the possibilities of reconstructing older forms of the language based on either internal or external evidence.) Deaf people in the United States are directed to the English language tradition through their formal education and become aware of the direct correspondences between English writing and spoken language. Signers have for many signs in ASL traditional names which are not necessarily the same as the glosses used by sign language linguists. Signers employ their names for signs in writing letters or in written telecommunication (via teletypewriter phone devices) to make the English more conversational or intimate. Linguists, on the other hand, tend to search for a unique gloss for each formationally distinct sign in order to separate those which may be shaped similarly but where the meanings are different. Sometimes, then, in deaf signers' usage, the same English word might be used for two signs which are nearly the same in form and where the context would allow the reader clues as to the particular sign intended. This is especially the case for signs which have no ready one-word English gloss. The linguist can view these difficult-to-gloss signs as revealing of relationships within the language which would not be apparent to most observers of sign language. We will now consider a case in which the metaphoric relationship between the language and the culture is pointed out by just such a conflict between ASL gloss and sign formation.

One issue which was raised at the beginning of this section was the question of expressive power. We noticed a number of lexical items which refer to what are usually called abstract ideas. In perhaps one way we avoided part of the issue about abstraction. When a language uses a three-dimensional

Figure 11: ASL COW

Figure 12: ASL SUB-
JECTIVE-TIME-PASSES-
SLOWLY

articulation space with two articulators, as sign
languages do, the representation of spatial relations between objects in the world is less a problem than it might be in some other modality. Relations such as "near," "above," "below," etc. may be
revealed through an overt lexical mention but are
often rather obvious from the placement in the
signing space of the objects partaking of these relations. What remains problematic for sign languages as for spoken languages is the specification
of temporal relations. ASL has a well developed
time word lexicon discussed in another place
(Frishberg & Gough 1973). One ASL sign in particular, however, can provide us with insight into
the complexity of lexical entries referring to temporal concepts. The sign is made by touching the
thumb of a Y-handshape (thumb and pinky only extended) to the temple and then moving the forearm
forward away from the face. The orientation tends
toward palm downward. The sign means something
like "subjective time passes slowly." If, for example, two people are working for two hours, one of
them might say at the end that they felt it must
have been longer than two hours. For that person,
subjective time passes slowly, and they could describe their perception using this sign. "We each
worked at our tasks for two hours but I felt that
it was much longer." The underlined portion of the
above sentence would all be expressed in ASL by a
single lexical item, not six, as in English. I
don't know any commonly accepted gloss for the
sign, but I am aware that some deaf people use the
English word cow to refer to the sign. The sign
COW is made in the same location with the same
handshape but with a different movement. COW
touches the Y-hand to the temple and nods the wrist
up and down (compare Figures 11 and 12) (10). Now,
what is crucially interesting about this example
for the present argument is not simply that there
exists such a complex idea all compactly expressed
in a single word in American Sign Language, although that is obviously interesting. Nor is the
interesting point only that the sign if often named
by the English word cow. But, if you will recall,
the overall argument is that there is a shared culture which is reflected in both ASL and in English
and that the same cultural images are brought to
the surface in the two languages in similar ways.
Consider the English word ruminating. Its literal
meaning refers to the chewing and swallowing

process of animals which chew cud; among those animals is of course the cow. The metaphoric meaning
is "ponder slowly" or "consider again and again."
We see that English makes a connection between slow
or careful thought and the actual chewing action of
cows and other animals. ASL evokes the notion of
perceived slow time in a sign which is named by the
English word <u>cow</u>. The actual sign COW is formationally quite similar to the target sign here SUB-
JECTIVE-TIME-PASSES-SLOWLY. The suggestion is that
the culturally shared notion which associates cows
with slow action, including slowed mental action,
surfaces in the two languages in comparable ways.
It is not the case that the sign in question here
means "ruminating," nor for that matter that <u>rumin-
ating</u> means "subjective time passes slowly," but
rather that the metaphoric extension of the attributes of cows in both languages covers similar semantic territory. The example here differs from all
of the previous ones in that it is not explainable
by borrowing from the oral language, but is a reflection of the cultural notions which are shared
by both languages.

<u>Writing or Other Visual-Graphic
Influence on Sign Languages</u>

We have already in several places begun to describe the relationship of written language (as a
representation of oral language) to signers. Deaf
people in situations that include formal institu-
tionalized education are usually taught to read and
write the oral language of the community. (They are
also taught speaking and lip reading.) ASL users
are instructed in schools for the deaf in English,
Chinese Sign Language users are taught Chinese, and
so on.

The written language, a part of the visual environment that surrounds the deaf person, also is
incorporated in intriguing ways into the sign language of the deaf community. The first — most obvious, and for our purposes least interesting — way
is through fingerspelling: American Sign Language
uses a one-handed manual alphabet as an adjunct to
the sign language system itself. The 26 letters
and the word AND are each represented by a handshape in a particular orientation. Rapid sequences of manual letter shapes can be used to

spell proper names or technical terms from English. Some relatively short words appear to be frozen in ASL as borrowed forms from English, occasionally with meaning shifts not predictable from the lexical item (Battison 1977).

Another part of written language is the set of symbols and punctuation marks. These, along with the letters and numbers, comprise the actively available orthographic character set for users of written English. Notice that parts of the character set are culturally specific. English and most other Roman derived scripts have quotation marks like this ",". Russian uses not only the Cyrillic alphabet but also some characters which don't occur in our symbol set, e.g., ≪ ≫ equivalent to quotation marks.

Consider the following two sets of non-alphanumeric characters listed below with their usual names in English:

Column I		Column II	
.	period	'	apostrophe
,	comma	___	underline
:	colon	$	dollar sign
;	semi-colon	¢	cents sign
∴	therefore	/	slash
−	hyphen	=	equals
+	positive, plus	÷	divide
−	negative, minus	#	crosshatch
%	percent	( )	parenthesis
" "	quotation marks	&	ampersand
?	question mark	°	degrees

These two lists are especially provocative in that there are no obvious criteria separating one from the other. Entries in the first appear to be as complex visually and as meaningful as those in the second column. Some of the symbols in each column have alternative pronunciations when read in normal English speech. For example, in 5 + 3 = 8 the + is read "plus," but in +5 it can be read either "plus (five)" or "positive (five)." Similarly, #

occurring before a digit will be read as "number," but occurring after a digit more likely would be read as "pounds" (<u>11</u>).

For users of American Sign Language the most important distinction between the two columns is that all of the examples in Column I have signed representations whereas all those in Column II do not. The English use and the ASL use of a particular symbol (whether in orthographic or manual representation) are not always equivalent. Some of the ASL signs have extended the meaning of a particular symbol or grammaticized it in a way quite different from the expected function.

Let's look at each of the entries in Column I and examine their signed counterparts. These six signs, PERIOD, COMMA, THEREFORE, SEMI-COLON, COLON, and HYPHEN, are made with index and thumb contacting one another slightly rounded from an otherwise compact fist. The sign PERIOD moves the hand sharply out from near the body and stops abruptly with the palm out, as if placing the dot on a vertical plane about 8 to 10 inches in front of the signer's body. COMMA likewise twists the outward-oriented hand in a movement so that the index-thumb juncture traces the shape of a comma. THERE-FORE makes three sharp movements like those used in PERIOD, first the upper and then the two lower dots of the familiar mathematical symbol, numbered here $\cdot_1$ $\cdot_2$ $\cdot_3$. SEMI-COLON, COLON, and HYPHEN are formed in analogous ways based on the same hand shape and movements composed of the elements described above.

The sign PERIOD occurs in much the same context as English speakers would use the word <u>period</u> to refer to the graphic symbol. The <u>Dictionary of American Sign Language</u> notes that PERIOD is used both for the "mark of punctuation...and the notion 'the end,' 'that's it'." There also exists a compound form which includes an initial element of touching the required handshape to the mouth. The same sign can be used to represent a decimal point in numerical expressions with decimal fractions. This usage is another indication that the sign does not translate the English word <u>period</u> but rather is a manual representation of the graphic symbol (<u>12</u>).

Since the sign THEREFORE is the only lexical

item in ASL with this meaning, it is used wherever
the meaning is appropriate. The several other
punctuation marks are used in fairly restricted
ways like their counterparts in <u>spoken</u> English.
That is, where you might expect an English speaker
to say the word <u>comma</u>, as in dictation or proofreading, a signer would use the corresponding sign.

Similarly, HYPHEN can be used in reading aloud
or in citing names, as well as in other contexts
where reference is made to written English type-
script.

The plus mark (+) POSITIVE is formed by the
two index fingers, non-dominant extended vertically
and dominant hand outside crossing the other horizontally. Not only can this sign indicate positive
numbers and the addition process (although the sign
ADD is also possible and may even be preferred in
this context), but it is extended metaphorically to
create phrases such as "positive attitude." Furthermore, the technical uses of <u>positive</u> in physics
and photography have been borrowed from English.

Minus sign (-) NEGATIVE appears in manual form
in ASL with a non-dominant flat hand facing outward
(away from the signer) and the dominant hand's index finger contacting the palm surface horizontally (11). NEGATIVE is allowed to indicate negative
numbers and, less frequently, subtraction. More
often the sign SUBTRACT/REMOVE will show subtraction. Furthermore, NEGATIVE has metaphorical uses
corresponding to those for POSITIVE, including the
technical senses in the sciences and photography.
POSITIVE and NEGATIVE can be used together in contrast to indicate "pros" and "cons."

PERCENT (%) is signed with the hand in an O
configuration, all the fingers rounded and touching
the thumb; the arm then moves horizontally right-
ward (for right-handed signers) and then a short
distance downward. The movement can be considered
imitative of the contiguous portion of the percent
sign %, although notice that the handshape remains
constant throughout the sign. The sign may be used
to mean either "percent" or "percentage" and appears both following numbers and without numbers as
do the English words. Notice that the symbol %
will not normally occur in a written text without a
number preceding it.

The last two items in Column I are more interesting in that their uses are not restricted to those predictable on the basis of English. QUOTES is a sign made like the gesture for quotation marks: index and middle fingers extended from each hand are held facing outward and curl downward simultaneously once (14). Not only does QUOTES (a) set off a direct quotation from the body of text, (b) precede a cited title of a literary or artistic work, and (c) act as scare or shudder quotes before an approximate or tentative term; but QUOTES serves several other purposes as well. The compound form IDIOM in ASL, meaning frozen phrase or set expression, is made by making the sign SENTENCE and following it immediately with QUOTES (15). QUOTES can also mean "topic" or "subject" as well as "title." QUOTES can inflect for location and direction which will give it a verbal meaning, the initial location indicating source argument. The derived sign EXCERPT with these modulations can mean even "plagiarize" or "pick someone's brain."

The sign QUESTION is clearly the most complex of all of these signs derived from orthographic symbols in English. It is formed by moving the outward extended index finger downward, curving the arm slightly as it moves. Alternately the finger itself bends as a more refined variant of the same movement. In this formation the sign signals a yes/no question and can precede (or, less frequently, follow) the question. It also serves as the verb form for "ask" in the meaning of "ask a question" (rather than "ask for/request"). Sometimes one will see the repeated tag-question variant of the sign QUESTION in which the arm remains stationary while the index finger bends downward several times rapidly. This form as a tag follows the sentence being questioned, but the same form may be used by itself as an initial utterance with the value "let me ask" or "I have a question for you," accompanied of course by the appropriate "intonational" cues of forward head and raised eyebrows. Notice in each case above the hand has been oriented outward, away from the signer's body. When the hand turns toward the signer and executes the same movement, the change in directionality indicates a change of subject-object relations, and the resultant form means "someone asked me." With appropriate eye gaze it may even mean "you ask me" (16).

The sign QUESTION, then, apparently etymologi-
cally derived from English orthography, is unlike
the previously cited items such as COLON, HYPHEN,
or PERIOD, in that QUESTION has the full semantic
and syntactic value of a verb in ASL. Rather than
showing up only where English orthography would
have a question mark, QUESTION has been grammaticized completely — it is in fact the only verb in
ASL with this particular meaning. Not only can
this sign inflect for person by changes in orientation as mentioned above, but it can also undergo
most other regular verbal deformations. It can, for
example, mark the plurality of the indirect object
either distributively by repeatedly curling the index finger while moving the arm horizontally from
left to right (for right-handed signers), or col-
lectively by bending once sharply while moving
rightwards. The first variant indicates "ask each
person/several people the same question"; the second means "ask all/a group a question." Furthermore, QUESTION can pluralize in another way so that
by using all the fingers simultaneously, oriented
outwards, one achieves the reading "ask many questions" or alternatively "ask many people/many of
you"; the sign by itself doesn't distinguish between plural direct object or indirect object.
Similarly, by turning the palm toward oneself and
using the QUESTION sign's movement with all the
fingers at once one indicates either "many people
asked me" or "he/they asked me many questions."

Several other signs play off of the same
orthographic unit. The repetition of the basic
sign QUESTION alternately by the two hands gives
the sign INTERROGATE, meaning "ask many successive
questions." A QUESTION made once sharply at the
forehead, palm oriented outward, gives the sign
PUZZLED (17). Turned toward the forehead and repeated (à la the tag question form), the sign becomes SUSPICIOUS. Lastly, by repeating the sign
simultaneously with both hands vertically downwards
and oriented outwards (optionally ending in flat
hands oriented palm down), we get the nominal sign
EXAM.

We have seen the apparently arbitrary division
between two categories of non-alphanumeric characters in English orthography. One set (listed in
Column I) has conventionalized ASL representations. Some of these signs go beyond the semantic

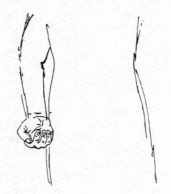

Figure 13:   Chinese Sign Language CANTON (= AdaSL
             WAIST-BEADS)

Figure 14:   China with Peking (P) and Canton (C)

value of the orthographic symbol in English, taking
on specialized meanings and regular productive
grammatical forms. The second set (Column II) of
characters have no sign representation in ASL. The
meaning they have for readers, however, can be
translated by signs which are in formation unrelated to the orthography; i.e., AND is not a manual
re-creation of the ampersand (&), nor is DOLLAR in
any of its several variants visually similar to $.
It seems possible that new signs could be created
to directly represent any of the non-alphanumeric
characters. In some of the recently devised sign
systems that are attempting to represent English
more directly, the apostrophe is shown in fingerspelled sequences by a twist of the wrist on the
letter following the mark. The usual cases of
course are 's and 't (possessive and contraction of
not). We would predict that a form like 'll would
not use this device but would instead have the
COMMA mark held slightly higher in the signing
space, followed by fingerspelled L-L. The other
symbols presumably would also have fairly consistent representations across signers but would be
expected only in the dictation or proofreading contexts. The meanings indicated by the orthography
are signed by forms that are unrelated to the written representation.

Beyond the incorporation of punctuation marks
into ASL we find in ASL and in other sign languages
the use of other graphic conventions. The map is a
fairly standard visual representation within a community. In formal educational settings for the
deaf, maps are apparently in widespread use. ASL
as a result has signs NORTH, SOUTH, EAST, and WEST,
which are simply initial letters N, S, E, and W
moved in straight lines up, down, and horizontally
to one side or the other. The conventionalization
of the orientation of a map thus has been borrowed
into ASL in the direction of movement in these
signs.

Chinese Sign Language has names for the five
cities in which there were deaf schools set up by
the British missionaries. While all the names are
interesting in formation, the one which is relevant
to the present discussion is CANTON. This sign is
made by touching the cupped hand held palm up to
the hip (see Figure 13). A glance at the map of
China (Figure 14, allowing for left-right rever-

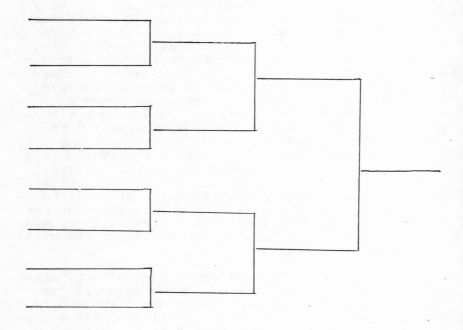

Figure 15:   Schema for an elimination tournament

sals) reveals this sign to be the representation of Canton's location with respect to the rest of the country. Considering traditional Chinese views that the center of the world is in Peking, one can easily understand how Canton ends up on the hip, at the lower edge of the signing space.

Another example from ASL is based on a standard-ized visual-graphic form. The sign ELIMINATION-TOURNAMENT is based on the chart in Figure 15. In the version shown, it has eight players or teams competing against one another, so that the winner of each trial meets the winner of one other trial, until one person or team is left. The sign begins with the fingers of both hands spread, pointing out-wards toward the addressee with dominant hand held over nondominant hand and moves the hands away from the signer successively, closing the fingers from four through to one and, finally, moves the domi-nant hand (now with the index only, pointing for-wards) from on top to in front of the nondominant hand. The speed and smoothness of the movement in this sign is obscured by the length and awkwardness of the verbal description (18).

The wide-ranging evidence offered above is in-tended to support the general argument of this pa-per; that is, the cultural setting of a particular language may have specific influences on the nature of lexical items in the lanugage, and thus we cannot expect sign languages, which exist in a given cul-tural setting, to be universally intelligible. The particular orthographic conventions which have been incorporated by signers of American Sign Language are often elements of print culture. These elements are shared among the deaf and hearing persons in the society, and have been adopted into ASL with speci-fic meanings often corresponding to their uses in print culture (;, :, ., etc.) but sometimes extend-ing from those uses that seem to be borrowed direct-ly (PERCENTAGE, QUESTION, etc.). Furthermore, other examples indicate the more general availability of all graphic elements (maps, charts, figures, etc.) to be incorporated in language in the visual mode. We would, however, hardly expect signers from non-print cultures to find such manual-visual represent-ations either immediately intelligible or motivated for themselves. It must follow, then, that the use of print culture symbols as the etymological source for signs in ASL or other sign languages (cf. CAN-

TON) is another demonstration of the local and cul-
turally-bound nature of sign languages of the deaf
(19)

Conclusions

In summary, we have seen how sign language va-
ries across geographic boundaries in much the same
way as spoken languages vary. We find structural
differences in sign language systems that prevent
mutual intelligibility among the different languages. In some cases, the repertoire of component
parts of signs (= words of a sign language) is
distinct from language to language. In other cases, even where the same component parts are shared
across languages, the systematic relationships between those parts are different. So, there is no
basis for the pervasive myth that "sign language"
is a universal communication system among deaf people, much as Esperanto, an artificial language, was
hoped to be among hearing people (20). Sign languages of the deaf are naturally occurring systems,
with different specific configurations of the hands
and other parameters depending on the particular
socio-cultural context.

Furthermore, we have seen how sign languages
may share certain meaning relationships with the
spoken languages of the culture they belong to. In
some cases, the shared meaning can be attribed to
borrowing but, in other cases, the shared meaning is
evoked by a cultural image which is not revealed in
the spoken language through overt morphology. The
non-universality of sign language representation is
thus connected to the more general variability of
culture around the world. A specific case showing
the relationship of American Sign Language to written English non-alphanumeric symbols was discussed
in greater detail, making the argument once again
that the shared culture (in this case, printed English) is represented in both languages. The examples really bring out how the signs are comparable
in usage and meaning to the printed symbols rather
than to the English words that translate the symbols. That is, the categories defined by the symbols of printing are used or extended in the sign
language of this culture as well as in the written
English culture. We do not, however, expect these
symbols to have representations in sign languages
from non-printing communities, nor would we even

predict these symbols to have identical representations in sign languages from cultures that share the printing conventions of our own. Rather, our study leads us to a greater awareness of the differences among different sign languages and to a greater appreciation of the diversity of sources for sign language representations.

Notes

1. Data from Kagobai's Rennell Island Sign Language are available in Kuschel (1973 & 1974), Providence Island in Woodward & de Santis (1975), Woodward, de Santis, & Washabaugh (1976), and Adamorobé Sign Language from my own field notes (Frishberg 1976b). Data for Chinese Sign Language are from Goodstadt (1970).

2. There is some indication (yet to be confirmed) that there is a higher percentage of left-handedness among the deaf in the United States. No careful large-scale study has been done of the population of signers as distinct from deaf people.

3. The signs of ASL will be indicated by English glosses written in all capitals. The English is to be taken as approximate rather than as definitive. There are many cases where several words in English are necessary to gloss a single sign. In this case, the gloss will be joined by hyphens, indicating a single sign, but multiple-word glosses joined by slashes (/) indicate alternative translations of a given sign. Signs in other sign languages are indicated in a similar way but are preceded by the language name or abbreviation (ASL, AdaSl, CSL, PISL).

4. This language, shared by the 40 to 50 deaf members of the community, is unrelated to any other known sign language. It should not be confused with Ghanaian Sign Language, a derivative of American Sign Language, used in the state schools for the deaf in Ghana.

5. It also indicates a fingerspelled W in ASL.

6. Or letter F.

7. The one place I have seen this handshape — ring finger and pinky extended — is in a joke.

The sarcastic form of APPLAUD/PRAISE can use either
this shape or pinky-only extended shape, either of
which denotes the diminutive of APPLAUD with the
further connotation of irony.

8. It is interesting to note that signers will
allow the use of a sign like CHAT, which literally
refers to casual manual conversation, for a comfortable prolonged interchange between two hearing people, or between a hearing and a deaf person. The
literal meanings of signs referring to language behavior, either made on the mouth or in the neutral
space (viz., SIGN-LANGUAGE, CHAT, etc.), appear not
to restrict the usage of the signs. So, as mentioned in the body of the text, the signs made on
the mouth can be used for verbal manual activity
and apparently the reverse is true as well; that is,
signs made with handshapes and movements clearly
referring to manual language behavior can be appropriately employed for talking about conversations
that occur exclusively in the auditory mode.

9. Not all of the signs made on the nose fit
the semantic category described here. Several eth-
nic appellations (GREEK, ROMAN), animal names (MOUSE,
RAT, FOX, WOLF), and other words (FLOWER, ABSORBING/
INTERESTING) use this location as well.

10. It is important to realize that the connection between COW and the sign referring to subjective
time is probably not an etymological one. COW is
from an older attested form that uses two hands and
suggests the cow's horns. SUBJECTIVE-TIME-PASSES-
SLOWLY is probably derived from the two signs FOR
and CONTINUE, although this etymology is not attested. Compare, for example, the compound FOREVER
(FOR + ALWAYS + STAY/CONTINUE).

11. Notice also that the symbol 's is never pro-
nounced as "apostrophe," except in metaorthographic
exchanges, such as proofreading: "You left out the
apostrophe after the word W." Similarly, # is not
usually read as "crosshatch," when it occurs in actual printed text. A number of these symbols signal
intonation changes in the vocal rendition of written
text.

12. PERIOD does not occur to make abbreviations.
The indication of separate letters in acronyms
shared by English and ASL is accomplished by the

fingerspelled handshapes taking on a circling movement for each letter, U.S.A., R.I.T. PERIOD does
not appear either in other abbreviations which occur
in both languages, e.g., A-P-T (not *A-P-T-PERIOD),
from apartment, M-A-S-S (not *M-A-S-S-PERIOD), from
Massachusetts, O-C-T (not *O-C-T-PERIOD), from
October, etc. Likewise, PERIOD does not appear in
ASL where English orthographic conventions prescribe
a period in the position following a digit in a
list, as: 1.
            2.
            3.
and, similarly, neither do we find PERIOD as part of
the formation of fingerspelled sequences which are
conventionalized abbreviations from English but
which English does not use as abbreviations, e.g.,
H-C, handicapped, H-S, high school, etc.

13. In formation, NEGATIVE resembles such signs
as the letter grades A-GRADE, ... E-GRADE, and ZERO,
all of which use the same base handshape, location,
and orientation as NEGATIVE. ZERO not only indicates a zero score in school or sports but can also
be the second element in several emphatic compounds:
SEE-ZERO, etc. See the Dictionary of American Sign
Language, p. 88 ff. for further notes on ZERO.

14. The gesture often repeats the motion twice.

15. The formation of SENTENCE in the compound
ends with the hands in the proper location for beginning the sign QUOTES, thus signaling word boundaries.

16. The variation in tense of English gloss
should not be taken as significant here. ASL does
not structure verb tenses in the same way as English
and, therefore, the glosses of signs are apparently
inconsistent on this point. The sign QUESTION,
when used as a verb, needs no overt pronoun markers,
although they often occur. In this respect, it acts
like a large class of verbs which allow orientation
or directionality (cf. Peng's discussion of Directionality in this volume) and changes to signify
shifts in subject-object relations. Other verbs
which behave like QUESTION include GIVE, SHOW, PAY,
ADVISE, ARREST, TAKE, IMPRESS, and TEACH.

17. This sign is particularly interesting in
that it apparently by coincidence is opposite in
meaning and form to the sign UNDERSTAND, which

flicks the index finger up from a fist while facing
the forehead.

   18. In fact, the signer can cite the winner simultaneously, given the type of tournament, by naming the champion in a separate sign and, then, signing ELEMINATION-TOURNAMENT with the additional final
movement of the other hand from behind to on top in
the bent 5 configuration facing palm downwards.
This configuration is that of the dominant hand in
the sign CHAMPION.

   19. Of course, the claims presented here make no
prediction about the learnability of ASL by signers
from other communities, despite the absence of
print culture from their environment. These signs
derived from graphic symbols would for speakers of
other languages be no less learnable but would be
totally arbitrary signs (in the sense of Peirce)
rather than the somewhat iconic formations they
seem to us.

   20. But see also World Federation of the Deaf
(1975) for an attempt at a universal international
sign language.

References Cited

Battison, Robbin M.
    1974 "Phonological Deletion in American Sign
          Language" Sign Language Studies   5.1-19.

Battison, Robbin M., Harry Markowicz, and James
Woodward
    1975 "A Good Rule of Thumb: Variable Phonology
          in American Sign Language" in F. Fasold
          and R. Shuy (eds.), Analyzing Variation
          in Language, pp. 291-302, Washington,
          D. C.: Georgetown University Press.

Bellugi, Ursula and Edward S. Klima
    1975 "Aspects of Sign Language and Its Struc-
          ture" in J. Kavanaugh and J. Cutting
          (eds.), The Role of Speech in Language,
          pp. 171-203, Cambridge: The M.I.T. Press.

Fischer, Susan D.
    1973 "Two Processes of Reduplication in the
          American Sign Language" Foundations of
          Language   9.469-80.
    1974 "Sign Language and Linguistic Universals"
          in Rohrer and Ruwet (eds.), Actes due
          Colloque Franco-Allemand de Grammaire
          Transformationelle, Tubingen: Max Nie-
          meyer Verlag.

Frishberg, Nancy
    1975 "Arbitrariness and Iconicity: HIstorical
          Change in American Sign Language" Lan-
          guage 51.696-719.

Frishberg, N. and B. Gough
    1973 "Time on Our Hands" paper presented at
          the Third Annual CLA Meeting, Stanford.

Goodstadt, Rose Y.
    1970 Speaking with Signs: A Sign Language
          Manual for Hong Kong's Deaf, Hong Kong:
          Hong Kong Government Printer.

Greenberg, Joseph H. (ed.)
    1966 Universals of Language, Cambridge: The
          M.I.T. Press, Second Edition.

Klima, Edward S. and Ursula Bellugi
    n.d. The Signs of Language, Cambridge:
          Harvard University Press.

Kuschel, Rolf
    1973   "The Silent Inventor: The Creation of a
            Sign Language by the Only Deaf-Mute on
            a Polynesian Island" Sign Language Stu-
            dies, 3.1-27.
    1974   A Lexicon of Signs from a Polynesian
            Outlier Island, Copenhaven: Psykologisk
            Laboratorium.

Lacy, Richard
    1974   "Putting Some of the Syntax back into
            Semantics" unpublished paper, La Jolla:
            UCSD.

Lane, H. L.
    1976   The Wild Boy of Aveyron, Cambridge:
            Harvard University Press.

Stokoe, William C.
    1960   Sign Language Structure: An Outline of
            the Visual Communication Systems of the
            American Deaf, Studies in Linguistics,
            Occasional Papers 8, Buffalo: University
            of Buffalo.
    1973   "Classification and Description of Sign
            Languages" in Thomas Sebeok (ed.), Cur-
            rent Trends in Linguistic Theory, Vol.
            12, The Hague: Mouton.

Stokoe, William C., Dorothy C. Casterline, and
Carl G. Croneberg
    1965   A Dictionary of American Sign Language
            on Linguistic Principles, Washington,
            D. C.: Gallaudet College Press (reprinted
            in 1976 by Linstock Press).

Ullmann, Stephen
    1966   "Semantic Universals" in J. H. Greenberg
            (ed.), Universals of Language, Cambridge:
            The M.I.T. Press, Second Edition.

Washabaugh, W., J. Woodward, and S. de Santis
    1976   "Providence Island Sign Language" paper
            presented at the 51st Annual Meeting of
            the Linguistic Society of American,
            Philadelphia.

Woodward, James C., S. de Santis, and W. Washabaugh
    1975   "Getting Back to Nature: Unhomogenized
            Linguistic Analysis" a revised version

of a paper presented at the Fourth Annual NWAVE Conference, October, Georgetown University.

World Federation of the Deaf
        1975   Gestuno: International Sign Language of the Deaf, Carlisle: British Deaf Association.

The American Sign Language Lexicon and Guidelines for the Standardization and Development of Technical Signs

3

Frank Caccamise, Richard Blasdell,
and Charles Bradley

Abstract: The recent rapid expansion in technical educational opportunities for the deaf has led to the need for a structured approach to the standardization and development of technical signs. A project designed to provide such a structured approach was established at the National Technical Institute for the Deaf (NTID) in September, 1975. This project has as its goal the establishment and maintenance of a technical sign bank which can serve the communication needs of all deaf students, their instructors, and interpreters. A primary element of this project is a set of guidelines based on our knowledge of the structural characteristics of the lexicon of American Sign Language. At NTID planning and initial data collection has begun on research designed to explore the relationship between visual physiology and the structural characteristics of signs which this set of guidelines describes.

Introduction

Of the two stated aims for this symposium this paper relates primarily to the first one - "...to account for certain characteristics that prevail in sign language." Specifically, this paper discusses a project for standardization and development of technical signs, with emphasis placed on a set of guidelines developed for use in this project. These guidelines, which are based on our knowledge of the American Sign Language

*This study was conducted in the course of an agreement with the Department of Health, Education, and Welfare.

(ASL) lexicon, describe the structural characteristics of signs which have evolved naturally.
These guidelines contribute to a needed preliminary step in achieving the first aim of this
symposium; that is, they identify or describe
characteristics that prevail in ASL, hence providing a base from which to explore our interest
in accounting for these characteristics. Using
this base, research designed to explore the
relationship between sign characteristics and
visual physiology has been planned at the National
Technical Institute for the Deaf (NTID).

Rationale for Technical Sign
Standardization and Development

It is recognized that a prescriptive approach
to standardizing and developing any language
or aspect of a language can lead to a decrease in
the ability of the language to adapt to needed
modifications, additions, etc. for purposes of
communication. However, recent rapid expansion of
technical educational opportunities for hearing-
impaired persons has precipitated the need for a
structured, but flexible, approach to technical
sign standardization and development. This need
is based on several factors. First, the rapid
expansion in technical educational opportunities
for the deaf has led to the need for deaf Americans to learn and use technical vocabulary which
few of them used in the past. Since this vocabulary was not previously needed, sign lexical
items for this vocabulary is largely nonexistent.
Second, interpreters at NTID have reported that
deaf students often have difficulty reading
fingerspelling, and that it is difficult for some
interpreters to keep pace with an instructor
giving a technical lecture when many of the words
need to be fingerspelled. Third, results of
studies at NTID have shown that deaf students
receive signs better than fingerspelling under
simultaneous manual-oral, interpreted manual-oral,
and manual-only test conditions (Caccamise and
Blasdell 1976). Fourth, and of foremost concern,
instructors having minimal knowledge of the linguistic structure of existing signs have begun to
invent sign equivalents for English technical
vocabulary, and some of these invented signs
violate the structural (cheremic or formational)

patterns of ASL signs which have evolved naturally, some are exactly like (or very similar to)
existing signs, and some are used for two or more
distinct technical concepts.

Based on the above facts it was decided that
a project designed to provide a more structured
approach to technical sign standardization and
development was needed. This project, which was
initiated at NTID during the Fall, 1975, has as
its primary goal the establishment and maintenance
of a technical sign bank which can serve the
technical communication needs of all deaf students, their instructors, and interpreters. The
remainder of this paper will concentrate on one
element of this project — the aforestated technical sign guidelines which are based on the
lexicon of ASL. Other elements of this project
are described in Caccamise et al., 1977.

Tentative Guidelines for Standardization and Development of Technical Signs

Stokoe et al. (1965), Battison (1974), and
Lane et al. (1976) have identified four major
parameters of ASL signs: (1) hand position; (2)
hand movement; (3) handshape or configuration; and
(4) hand orientation, which primarily refers to the
direction of the palm. The guidelines for the
technical sign project at NTID are based on
existing ways in which these four sign parameters
are combined in ASL signs. In application,
therefore, these guidelines may be used to assess
signs in terms of acceptable and unacceptable
combinations of the four major parameters of ASL
signs. Since one of the elements of language
planning is to ensure stability, it follows that
sign standardization and development should
follow a set of guidelines based on the structure
of signs which have evolved over time, and which
have withstood the test of time and usage. The
ASL lexicon consists of such signs. The logical
consequence of not following such guidelines is
that signers themselves will change the form of
invented signs to be compatible with the more
naturally developed signs of ASL. An example of
this is discussed in the last paragraph in the
section of this paper entitled "Some Preliminary
Final Thoughts."

As indicated by the title of this section, the guidelines developed at NTID are tentative and are expected to undergo further refinement as more is learned about manual communication in general, and the structural patterns of signs within the ASL lexicon in particular. The guidelines developed to date at NTID for technical sign standardization and development are as follows:

1. The Signing Space. Signs generally fall within a particular signing space, bounded by the top of the head and the area just above the waist, with the space towards the sides of the body involving a comfortable, but not fully extended, reach of the arms. The physical center of the signing space is the hollow of the neck (Frishberg and Gough 1973). It is important to recognize that for a variety of reasons signs may be made outside of this general signing space (e.g., for theatrical signing, for emphasis, etc.).

The mouth area is considered not to be within the general signing space. This is partly based on the fact that it has been consistently observed, and reported by signers themselves, that when reading signs people tend to watch the face area of the signer, rather than the signer's hands per se. This agrees with the contention that facial expressions and lip movement (whether or not words are mouthed or spoken) are important parts of manual communication. Also, in making observations of historical changes in signs, Frishberg (1975) found that signs made in the mouth area tended to displace away from the mouth toward the sides of the mouth or the chin area. For example, the sign RED used to be made on the lips (and is still depicted that way in many sign books), however, many signers now make the sign lower, on the chin. Therefore, an effort should be made not to obstruct the mouth area when signing (also, see guideline #2).

2. Signs and the Visual Center of the Signing Space. The visual center of the sining space is the nose-mouth area (Siple n.d.). Many signs are made in close proximity to this visual center, but seldom are signs made within this center.

In addition to what is presented in guideline #1, the logic for this is based on the following:

A. Visual acuity is sharpest near the visual

center of the signing space, becoming poorer as you move toward the periphery or away from this visual center (Siple n.d.).

B. Research has shown hearing-impaired persons to perform better on tests of receptive communication skills when oral and manual communication modes are used together as opposed to either alone (Johnson 1975; Klopping 1971; Stuckless 1975). If lip movement is contributing to this superiority of the combined oral-manual condition over either oral or manual alone, then, blockage of the mouth area may reduce efficiency of communication.

3. The Relationship of Sign Hand Position and Other Sign Parameters. Signs made near the visual center of the signing space tend to have finer distinctions for all parameters than those signs made in the periphery (Siple n.d.). The probability of perceiving detailed information is greatest in areas of high acuity, and areas like the face which have a large number of visually distinguishable landmarks. Therefore, smaller motions and distinctions among signs is feasible in such areas (e.g., areas close to the visual center) as opposed to lower acuity areas (e.g., areas more peripheral to the visual center).

4. One-hand and Two-hand Signs. In general, signs made near the visual center of the signing space involve the use of one hand, while signs made near the periphery tend to involve two hands having symmetrical handshapes, movements, positions, and orientations.

It is logical to expect that duplication of information through the use of two symmetrical handshapes, etc., is of greater importance for signs made in peripheral or low acuity areas, as opposed to signs made closer to the high acuity visual center area of signing. In fact, duplication involving the two hands in the high acuity face area is often overredundant (Frishberg 1975). Based on these facts, it is suggested that:

A. For signs made in the neck and face area use one hand, with the position toward the periphery for face area signs (also, see guideline #2).

B. For signs made below the neck, use two hands having the same handshape. Also, the movements, positions, and orientations of the two hands should be symmetrical (Battison 1974).

Movements are considered symmetrical if the hands have the same basic movement either in the same or opposite directions.

Positions are considered symmetrical if the hands contact the same position or corresponding positions on halves of the body.

Orientations are considered symmetrical if the orientations of the hands are the same or are polar opposites (i.e., reciprocals).

5. Guidelines Related to Two-hand Signs in which only One Hand moves.

A. The non-moving (passive) hand should have one of the seven neutral handshapes (1-A-S-B-C-5-0), or should have the same handshape as the moving (active) hand (Battison 1974).

B. When two or more such signs differ only in the handshape of one hand, this difference should occur in the moving hand. Logically, one would expect that for signs involving one moving and one still hand the receiver will attend to the moving hand.

6. Number of Handshapes per Sign. Most signs in ASL (whether one-handed or two-handed) require only one handshape on each hand. However, some signs require that the handshape changes during the articulation of the sign. For example, MILK is made in neutral space with one hand, and that hand closes from a C handshape to an S handshape (repeatedly); the sign SPAT changes repeatedly from 0 to B handshapes; PRINT uses two handshapes, one with the thumb and forefinger separated, and one in which they contact. What is important to note about these handshape-changing signs is that they all involve no more than two handshapes; there is no ASL sign which uses three or more handshapes. Undoubtedly, this has naturally evolved so as to limit the number of handshape discriminations found in the native ASL lexicon. In view of such an absolute restriction found in the native ASL lexicon, it is recommended that no invented signs use more than two handshapes.

7. Guideline for Signs Involving Contact. Four
major areas of "contact" in signs are the head,
trunk, arm, and hand. Battison, Markowicz, and
Woodward (1975) found that ASL signs are systematic in regard to the contacts in these areas, and
this adds to the redundancy factor in signing which
is necessary for efficient language reception.

Signs made with double contacts are made within
the same major area (e.g., INDIAN has both contacts
on the head, WE has both contacts on the trunk,
etc.). The exceptions to this rule are signs which
are historically derived from compounds, and move
from a contact in one major area to a second contact in another major area (e.g., REMEMBER, a compound derived from THINK + SEAL, contacts first the
head area and, then, the opposite hand; DAUGHTER,
derived from GIRL + BABY, contacts first the head
and, then, the arm).

It is, therefore, recommended that signs having two contacts have both contacts within one of
the four major areas previously listed; that is, the
head, trunk, arm, or hand.

8. Semantically Related Signs. Signs which
are semantically related (that is, related in meaning) are often related in terms of formation. "Semantically related" refers to those signs whose
English glosses are approximately synonymous with
each other, and/or whose English glosses have a conceptual relationship. For example, a change in
handshape can change the meaning of the sign GROUP
to CLASS, ASSOCIATION, or SOCIETY. All of these
are semantically related in that each refers to a
group or a kind of group. Also, all are formationally related in that their corresponding signs have
the same position, movement, and orientation (cf.
Sign Symbolism discussed by Peng in this volume)
with only handshapes distinguishing among them.
This relationship between semantics (meaning) and
formation of signs is an example of internal
structural at the morphological level (Frishberg,
1975). This type of structural relationship
should be taken into account by those involved in
sign standardization and development.

Table 1 provides a summary listing of these guidelines.

TABLE 1

A SUMMARY LISTING OF TENTATIVE GUIDELINES FOR
STANDARDIZATION AND DEVELOPMENT OF
TECHNICAL SIGNS

1. THE SIGNING SPACE

2. SIGNS AND THE VISUAL CENTER OF THE SIGNING

3. THE RELATIONSHIP OF SIGN HAND POSITION AND
    OTHER SIGN PARAMETERS

4. ONE-HAND AND TWO-HAND SIGNS AND SYMMETRY

5. GUIDELINES RELATED TO TWO-HANDED SIGNS IN
    WHICH ONLY ONE HAND MOVES

6. NUMBER OF HANDSHAPES PER SIGN

7. GUIDELINE FOR SIGNS INVOLVING CONTACT

8. SEMANTICALLY RELATED SIGNS

Accounting for Guidelines

As previously stated, the guidelines listed
are based on our knowledge of signs within the
lexicon of ASL. They provide a description of the
characteristics that prevail among signs, and
provide some explanation as to why these characteristics do prevail.

In attempting to account for sign characteristics it is logical to assume that the physiology of those body parts involved in expressing
signs will necessarily play a significant role in
determining the possible characteristics of signs

(see Battison 1974, for a discussion of this).
An analogous situation exists in speech production
which is necessarily governed by the physiological
capabilities of the vocal system. Also, some of
this accounting for sign characteristics may be
expected to be derived from an understanding of
the physiology of the visual system (see guidelines 1 through 5 especially); since signing
is a gestural expressive communication system
which depends on the eyes for reception, senders
of signs may be expected to produce signs in a
manner which will be consistent with the visual
capabilities of the receivers. Again, an analogous situation exists for oral or speech communication where the primary energy in the expression of human speech matches the most sensitive
frequency range of the human ear for reception of
auditory stimuli (500 Hz-3000 Hz).

At NTID we are concentrating our research
efforts on the functioning of the visual system in
communication reception involving manual communication (signing and fingerspelling). This
research is designed to provide information
needed in order to better understand and account
for the sign characteristics described in the
guidelines listed in this paper.

The major research project underway at NTID
involves monitoring of eye movements during
presentation of signs, fingerspelling, and speech.
If lip movement is important in communication
involving the use of signs and the nose-mouth area
is the visual center of the signing space (as
suggested by guidelines #1 and #2), then, monitoring of eye movements during the presentation
of signs and speech should show the eyes to
primarily focus on the nose-mouth area regardless
of the position of the sign. If this is true, then,
support is provided for recommendations in guidelines #1 and #2 that signs should be made in close
proximity to the visual center of the signing
space, but they should not block the nose-mouth
area.

In addition to monitoring of eye movements,
the above described study will include measures of
comprehension by subjects. A comparison of
primary position of eye focus and comprehension
performance may provide us a basis for

instructional strategies with students. That is,
should students be instructed to learn to focus on
the hands directly, or should they be instructed
to focus on the nose-mouth area while using their
peripheral vision for signs and fingerspelling?
Our conjecture is that during combined signs and
speech communication receivers will tend to focus
on the nose-mouth area in order to pick up the
finer speech and non-speech movements of the lips,
with peripheral vision being used to pick up the
larger, more gross parameters or characteristics
of signs. However, during combined fingerspelling
and speech there may be a tendency to focus more
on the hands in order to recognize the finer motor
movements and other distinctions needed to effec-
tively receive fingerspelling. Support for this
has been provided by Reich (1976), who found that
for combined presentation of speech and fingerspelling, highest reception scores occurred under
the condition of closest proximity between the
hand and face.

A second project being planned relates to
guideline #5 which suggests that for signs involving one moving and one still hand the receiver
will attend to the moving hand. This second
project will involve the presentation of two-
handed signs in which only one hand moves. The
moving hand will be varied, sometimes being the
left hand and sometimes the right. If guideline
#5B is correct the eyes should follow the moving
hand, regardless of whether the moving hand is the
left or right hand.

Many other research projects related to our
guidelines are possible. We would like to leave
these to the imaginations of our readers, and
welcome their assistance in providing further data
as to why certain characteristics do prevail in
sign language.

Some Preliminary Final Thoughts

In any project involving language planning,
sensitivity to the input, needs, and possible
misunderstanding of others is essential. Therefore, we wish to make a few brief statements to
better clarify our approach to technical sign
standardization and development.

First, the guidelines discussed in this paper are not only tentative, but they are in fact guidelines and not strict rules. Usage by deaf people and native signers is the prime determiner as to whether a sign is acceptable or unacceptable.

Second, sign standardization and development does not mean there is a need for one sign and only one sign for each technical concept or referent. Rather, similar to the strength and flexibility provided by synonyms in English, the existence of more than one sign for some technical concepts or referents may provide the same strength and flexibility needed for efficient transmission of information when manual communication is part of a communication event. Preliminary analysis of signs collected at NTID indicates that for a number of technical concepts there exists more than one acceptable sign. If further investigation shows that these signs are equally acceptable to instructors, students, and interpreters, then each of these signs will be recommended as appropriate for use in the classroom.

Third, to date the project at NTID has involved only the collection and assessment of existing technical signs. Although the invention of new signs may be deemed appropriate at some future time, it is believed that a thorough collection and assessment of existing technical signs should be undertaken first.

Fourth, the use of fingerspelling is a viable alternative to sign invention.

Fifth, the project described in this paper is not an attempt to standardize a sign language or sign system since it does not include any considerations of language except for single lexical items.

Sixth, the Technical Sign Standardization and Development Project at NTID is not considered an isolated effort. Rather, it is a project with national implications, and which therefore must take into account the needs, efforts, and beliefs of others concerned with better meeting the communication needs of deaf persons and those of us who have the opportunity to interact with deaf

persons. Input is being solicited from other
programs involved in and/or concerned with sign
standardization and development. This input
should help us to provide support for, and in-
corporate into our efforts, the work of these
programs.

Seventh, the guidelines discussed in this
paper are based exclusively on the American Sign
Language lexicon. The extent to which other
natural sign languages may reflect these guidelines has not been explored.

Eighth, although structured research is
needed, the importance of observations of natural language situations may be expected to
continue to provide valuable input about the
characteristics of signs and signing. For
example, observational support for guideline #5B
has been provided by two authors of one of the
new invented sign systems designed to represent
English. In their sign book these authors
listed invented two-handed signs with one hand
moving for CAR, TRUCK, and BUS. Although the
only difference among the three signs in their
book was the handshape of the non-moving hand
(violation of guideline #5B), when asked how
they signed these three words the two authors
placed the distinguishing handshape on the
moving hand.

Summary

A project on technical sign standardization
and development has been discussed. Emphasis
was placed on a set of tentative technical sign
guidelines. These guidelines are based on the
structural parameters of the American Sign Language lexicon, and they provide a description of
the characteristics that prevail among signs.
In general, these guidelines describe acceptable
and unacceptable combinations of the four basic
ASL sign parameters: (1) hand position; (2) hand
movement; (3) handshape or configuration; and (4)
hand orientation. Research designed to provide
data needed to better understand why certain
characteristics prevail among signs is being
conducted at NTID.

Notes

The authors wish to thank the following persons who have served on the NTID Technical Sign Standardization and Development Project: Robert Ayers, Robbin Battison, Alice Beardsley, Anna Braddock, Karen Finch, Loy Golladay, Tracy Hurwitz, Richard Nowell, Paul Menkis, Paul Peterson, Pat Siple, James Stangarone, and Katherine Warren.

References Cited

Battison, R.
       1974 "Phonological Deletion in American Sign Language" Sign Language Studies 5.1-19.

Battison, R., H. Markowicz, and J. Woodward
       1975 "A Good Rule of Thumb: Variable Phonology in American Sign Language" in R. Shuy and R. Fasold (eds.), New Ways of Analyzing Variation in English, No. 2, Georgetown: Georgetown University Press.

Bradley, C., R. Ayers, K. Finch, and D. Kolakawski
       1976 "A Recording System for Signs based on Stokoe Notation" in Working Paper, Rochester: National Technical Institute for the Deaf.

Caccamise, F. and R. Blasdell
       1976 "Reception of Signs and Fingerspelling under Manual, Intepreted, and Simultaneous Test Conditions" in Working Paper, Rochester: National Technical Institute for the Deaf.

Caccamise, F. C. Bradley, R. Battison, R. Blasdell, K. Warren, and T. Hurwitz
       1977 "A Project for Standardization and Development of Technical Signs" American Annals of the Deaf 122(1).44-9.

Frishberg, Nancy
       1975 "Arbitrariness and Iconicity: Historical Change in American Sign Language" Language 51(3).696-719.

Frishberg, N. and B. Gough

1973   "Time on Our Hands" paper presented at the 3rd Annual California Linguistics Association Meeting, Stanford University.

Jacobs, J.
1975   "The Standardization of Signs Effort in Texas Schools for the Deaf" in Working Paper, Statewide Project for the Deaf, Austin: Texas School for the Deaf.

Johnson, D.
1975   "Communication Characteristics of NTID Students" Journal of the Academy of Rehabilitative Audiology 8(1&2).17-32.

Klopping, H.
1971   "Language Understanding of Deaf Students under Three Auditory-visual Stimulus Conditions" unpublished Ed.D. Thesis, University of Arizona.

Lane, H., P. Boyes-Braem, and U. Bellugi
1976   "Preliminaries to a Distinctive Feature Analysis of Handshapes in American Sign Language" Cognitive Psychology 8.263-89.

Reich, P.
1976   "Variables Affecting the Comprehension of Visible English" in Working Paper, Toronto: University of Toronto.

Siple, P.
n.d.   "Visual Constraints for Sign Language Communication" in E. S. Klima and U. Bellugi (eds.), The Signs of Language, Cambridge: Harvard University Press.

Stokoe, W., D. Casterline, and C. Croneberg
1965   A Dictionary of American Sign Language on Linguistic Principles, Washington, D. C.: Gallaudet College Press. (Reissued by Linstok Press, Silver Spring, 1976).

Stuckless, R.
1975   "An Interpretive Review of Research on Manual Communication in the Education of Deaf Children: Language Development and Information Transmission" paper presented at a Seminar on the Place of

Manual Communication in Education of
Deaf Children, sponsored by the Royal
Institute for the Deaf, London, England,
April 11-14.

4

Language Acquisition in Apes and Children

Lyn W. Miles

The continuing success of projects investigating the linguistic and cognitive abilities of apes has raised the issue of how the acquisition and use of language-like skills by chimpanzees and a gorilla relates to the ontogenic and evolutionary development of language in humans. Although these investigators have been careful not to claim that apes using sign language (B. Gardner and R. Gardner 1971; Fouts 1973) plastic chips (Premack 1971) or computer language (Rumbaugh and Gill 1976) were acquiring the same language facility as humans, their results have nevertheless challenged long held assumptions of human uniqueness. Based on the success of these projects, Rumbaugh and Gill (1976) suggest that some cognitive skills underlying language in humans may be shared by apes and that these abilities can be expressed linguistically when apes are given a suitable means of communication which avoids the limitations of their vocal apparatus. Now that several apes have been trained to use signs or other means of communication with humans it is possible to obtain enough data to begin to compare the linguistic and cognitive abilities of apes and children.

Comparisons between apes and children are dependent not only upon progress in ape language research, but also upon the methodological and theoretical perspectives of child language studies. Except for a few isolated early studies, the systematic investigation of human language acquisition is relatively recent. Investigators of child language have searched for an appropriate

paradigm with which to understand language development and recent work in the area has led to some
significant changes in the theory of how human
children learn language. The primary problem in
the study of child language has been how to con-
ceptualize the acquisition process itself. The
first attempts to characterize child language
acquisition were primarily descriptive. Using
large samples of children's speech these studies
described the number of words used or the age when
a particular construction was first used by groups
of children. Language acquisition was understood
as consisting of developmental milestones which
were similar for all children. Later, attention
was focused on how children come to understand the
structure of language and develop grammar. More
recently investigators have concentrated on the
cognitive abilities underlying language acquisition in the form of semantic relations. As an
extension of this approach, several investigators
have come to recognize the importance of the communicative functions of language. Language, they
argue, develops not in isolation, but within a
social context involving a variety of social and
cognitive processes. This most recent development
has important implications for understanding the
development of language in early hominids and
assessing the implications of comparisons between
apes and children.

It is not possible to directly compare the
use of plastic tokens by Sarah and computer language by Lana with normal language acquisition in
human children. Children do not normally acquire
language through these means, although there are
several experimental projects using these means
with retarded children in institutional settings.
In addition, Brown (1973) has urged caution in
interpreting Sarah's use of her plastic tokens on
methodological grounds and also because she did
not appear to use the tokens intentionally as a
means of communication. However, it will be somewhat easier to compare Lana's use of a computer
language with at least retarded children because
Lana's acquisition procedures have been carefully
controlled and also because Lana initiates communications through her computer keyboard.

It is much easier to compare the acquisition
of sign language by apes and children. Sign

language is already a language used by a large
community of humans, and a few chimpanzees have
learned sign language from birth like human children. /These apes and children can be compared on
the basis of how they acquire signs, the size of
their vocabulary, the way in which they combine
their signs to form sentences, the kinds of errors
they make, the signs they invent, and the way they
use signs to communicate. And then both deaf
children and apes learning sign can be compared
with hearing children learning speech. /

Sign Language Acquisition

Project Washoe was the first attempt to teach
sign language to an ape. Washoe was almost 1 year
old when her sign language training began and
after a little over 4 years of training she
learned 132 signs based on stringent acquisition
criteria (B. Gardner and R. Gardner 1971). She
began combining signs after only 10 months of
training when she was 2 years old. Like human
children (Klima and Bellugi 1970) she gradually
began to increase the length of her signed communications. Based on their experiences with
Washoe, the Gardners (R. Gardner and B. Gardner
1975) began a new project with 2 chimpanzees, Moja
and Pili who (in contrast to Washoe) were immersed
in a sign language environment from birth and were
taught signs by fluent signers including deaf
persons. Under these conditions both Moja and
Pili began to sign in their 3rd month, and Moja
began to use 2 sign combinations at the age of 6
months (R. Gardner and B. Gardner 1975). Roger
Fouts (1973) also found that a chimpanzee who was
given sign language training almost from birth
acquired her first sign at 4 months of age, and
Patterson (n.d.) reports that after 4 months of
sign language training the gorilla Koko began using sign combinations at the age of 14 months.

Braine (1963) and Brown (1973) report that
human children begin to form 2 word combinations
by approximately 2 years of age. While the age
at which these chimpanzees produce their first
signs and combinations seems early compared to
children's earliest speech, Schlesinger and
Meadow (1973), Mindel and Vernon (1971) and Stokoe
(1976) report that deaf and hearing children of
deaf parents acquire their first signs at 5 and

6 months of age and combine signs at about 14
months, approximately 6 to 10 months earlier than
speaking children combine words. Thus, the acquisition of signs and use of combinations by apes
is very similar to the first use of signs by
human children, especially when the apes are
trained from birth under typical human rearing
conditions. The Gardners (R. Gardner and B.
Gardner 1975) suggest that the earliest production of signs by apes and human children may be
because signs are easier to perform than words or
because it may be easier to recognize and reinforce gestural approximations than the early sounds
of speech.

There are a number of other similarities in
sign acquisition shared by apes and children. For
example, both apes and human children occasionally
invent signs. Bellugi and Klima (1976) report
that deaf children invent signs which are mimetic
(nonce signs or neologisms) yet also conform to
the systematic formal constraints of American Sign
Language. Several of the apes who have received
sign language training have also invented signs.
Washoe invented a sign for "bib," (B. Gardner and
R. Gardner 1971) and Koko invented signs for
"stethoscope," "bite," "tickle," and "note"
(Patterson n.d.).

Human children also must learn to generalize
their use of signs and words to novel objects and
situations (Clark 1973). The Gardners (B. Gardner
and R. Gardner 1971) and Fouts (1973) and other
ape language acquisition investigators have demonstrated that apes also learn to generalize their
signs. And like human children (Brown 1973), apes
also engage in babbling and self-signing. Thus,
the evidence concerning sign acquisiton suggests
strong similarities between apes and children in
both the rates and means of acquisition.

Sign Performance

Sign errors are of particular interest in
comparing sign performance in apes and children.
Bellugi and Klima (1976) report that errors in
sign language are not related to semantics or to
the iconicity of signs. Instead, sign errors
usually involve a variation in one of the systematic formational aspects of a sign, it's place of

articulation, hand configuration or significant movement.

During a study of chimpanzee sign language conversations, Ally and Booee made several sign form errors which conform to the human pattern. For example, when signing the name "Joe," Ally occasionally used his index finger instead of his little finger for his hand configuration. Booee also varied the place of articulation of especially the "tickle" sign (Miles 1977). Patterson (n.d.) reports that Koko also varied the place of articulation of several signs in her repertoire and that these variations seemed to be associated with slight differences in meaning. Bellugi and Klima (1976) report that human signers also affect the meanings of their signs through variations in the movement or place of articulation of a sign.

During the conversation study, Booee and Ally also occasionally performed signs with both hands simultaneously, or rapidly repeated signs. This has also been reported for Koko (Patterson n.d.) and deaf children (Hoffmeister et al. n.d.). Immediate sign repetitions occur more frequently in sign language than spoken English and can signal changes in meaning which have syntactic consequences (Fischer 1973). However, for very young deaf children it is not clear whether repetitions represent modulations in meaning or merely emphasis. Since immediate sign repetitions were rarely used by the human trainers during the conversation study the significance of repetitions in the apes' sign construction cannot be determined.

During the conversations, Booee also sometimes collapsed 2 signs into a contraction involving one motion. For example, Booee would sign "tickle there" by forming the "tickle" sign directly on the location of the desired tickle, usually his thigh, thereby changing the place of articulation while retaining the hand configuration and movement of the "tickle" sign. Patterson (n.d.) reports that Koko has also varied the place of articulation of her signs.

During the conversations, both Booee and Ally also used gestural intonation markers, such as holding their hands in position at the end of a

phrase to ask a question. On one occasion, Ally
asked, "shoe that box?" when he was asking the
name for a large rubber boot, and on another occasion Booee asked, "shoe tickle Booee?" when he
was trying to clarify a tickle game. /Washoe (B.
Gardner and R. Gardner 1971) and Koko (Patterson
n.d.) as well as other signing apes have also
learned to use gestural intonation to mark questions./ Children learning sign or spoken English
also use vocal or gestural intonation to mark
questions and only later learn to utilize inter-
rogatives (Bowerman 1973; Brown 1973).

Mean Length of Utterance

Brown (1973) has suggested that the acquisition of speech by human children can be divided
into 5 stages each with a Mean Length of Utterance
and an Upper Bound or longest utterance, based on
a count of morphemes. Each of these stages is
characterized by increasingly complicated linguistic operations. For children in the earliest
stage (Stage I) of language acquisition (2 to 3
year olds), Brown reports a Mean Length of Utterance target value of 1.75 and an Upper Bound
target value of 5. Calculating the comparable
Mean Length of Utterance for sign language is
complicated by the fact that sign language does
not modulate meanings by the addition of morphemes or the use of sequential devices such as
word order. However, Hoffmeister et al. (n.d.) has
suggested guidelines for applying Mean Length of
Utterance measures to the earliest stages of sign
acquisition which have been adapted for use with
apes (Miles 1977). Based on these measures, apes
exhibit a Mean Length of Utterance and Upper
Bound in the range of Stage I language acquisition
of human children. Ally showed a Mean Length of
Utterance of 1.66 with an Upper Bound of 4 and
Booee showed a Mean Length of Utterance of 1.79
with an Upper Bound of 6. Patterson (n.d.)
reports a Mean Length of Utterance of 1.82 and an
Upper Bound of 11 for Koko. Large Upper Bound
measures for Stage I acquisition usually indicates
the stringing of sequences of signs or words which
is common for both apes (Miles 1977) and human
children (Brown 1973). Stringing is not considered indicative of sentence formation complexity and thus does not alter the language
acquisition state classification.

Grammar

In the 1960's, many investigators of child language began to focus on language structure and to attempt to characterize the first grammar of children. This new focus was largely a result of Chomsky's assertion that children do not learn a repertoire of sentences, but instead acquire a set of rules which is capable of generating an infinite number of sentences. This approach focused on a description of the rules which children use at various stages to generate their utterances.

Initial attempts to characterize the development of grammar in children focused on the telegraphic nature of a child's speech and on word frequency distributions (Brown and Fraser 1963; Brown and Bellugi 1964). Other approaches (Braine 1963; Miller and Ervin 1964) defined 2 classes of words, pivot words and open words, which children used in different ways to construct sentences. Telegraphic analyses and pivot grammars did not adequately describe children's utterances, however, and especially failed to provide a basis for interpreting the different meanings of a word sequence.

The concern for constructing a first grammar for children was reflected in studies of language acquisition in apes which concentrated on sign order (B. Gardner and R. Gardner 1974, 1975; R. Gardner and B. Gardner 1974, 1975; Brown 1973). Since the grammar of Lana's computer language was highly constrained, Rumbaugh (1973) was able to demonstrate that a chimpanzee can construct a linguistic proposition which shows evidence of rule following behavior and can also recognize what constitutes a grammatical syntactic construction. In American Sign Language there is less reliance on sign order to express grammatical relations than in Yerkish, Lana's computer language, or English. However, regularities in sign order have been reported for apes. Fouts (1974) reported some regularities in Washoe's 2 and 3 sign combinations and Patterson (n.d.) has reported similar regularities for Koko's sign constructions. In the conversation study, both Ally and Booee showed some word order preferences (Miles 1977). Ally showed a preference for a noun to be preceded by the demonstrative, e.g., "that

ball," in 92% of his 2 sign constructions involving demonstratives. He also showed a subject-
verb-object sign order preference in 89% of his 3
sign constructions. But, this may be due to
Ally's extensive exposure to spoken English word
order and a subject-verb-object word order preference in the constructions of his human conversation partners. In American Sign Language fluent
signers are much more likely to use hand movements
and facial expressions to express modulations of
meaning. Deaf signers usually maintain a fairly
free sign order in their constructions (Bellugi
and Klima 1976) although deaf children exposed to
English may also show similar sign order preferences.

Apes and human children also sometimes perform signs simultaneously sometimes using both
hands at the same time. This has also been reported for Koko (Patterson, n.d.). Thus, although
there is no evidence to suggest that apes and
children differ in their early sign orders, the
characteristics of sign language and the limitations of grammatical analyses in general make
comparisons between apes and children using a
grammatical paradigm not very useful in the early
stages of language acquisition.

Semantic Relations

Because of the problems associated with grammatical approaches, more recent theories of language development (Fillmore 1968; Chafe 1970;
Brown 1973) have included semantics, or an analysis of meaning, as a basis for structural consideration. Bloom (1970) and Schlesinger (1971)
have argued that even very young children have
semantic intentions which they gradually learn to
encode into appropriate syntactic forms. Bloom
(1974) has shown that when a child says, "mommy
sock" it could mean different things depending
upon the context in which it was uttered. "Mommy
sock" uttered when pointing to the mother's sock
means possession or ownership, but when uttered
while the mother puts a sock on the child, refers
to an action with an object. Brown (1973) reports
that 8 prevalent semantic relations account for
70% of the relaxed conversation of a child in
Stage I. These relations include combinations of
agent, action, object, locative, entity,

possession, and demonstrative in 2 and 3 term
construction. Klima and Bellugi (1972) and
Schlesinger and Meadow (1971) report similar
results for deaf children.

In comparison, 68% of Booee's signed communications and 75% of Ally's signed communications expressed these same relations during relaxed conversations (Miles 1977). Using a
slightly expanded set of semantic categories 78%
of Washoe's sign combinations (B. Gardner and R.
Gardner 1971) and 75% of Koko's combinations
(Patterson n.d.) also express these prevalent
relations. Thus, several apes and deaf and hearing children use the same frequency of these
semantic relations in the earliest stage of language acquisition.

Communicative Competence

In addition to the structure of children's
early language, investigators have recently turned
to an examination of the function of language,
that of communication. Campbell and Wales (1970)
and Hymes (1971) have argued that it is also
important to study the development of communicative competence as part of the process of language
acquisition. Ryan (1974) has argued that both
grammatical and semantic analyses of children's
language have failed to account for the fact that
the form of an utterance is related to the communicative intention which motivates it. From the
very first, language is learned by children within
a system of social interaction. In learning to
request, greet, protest, make reference, identify,
etc., children learn both the social conventions
of communication and its special human form:
language (Lieven and McShane 1976).

The analysis of communicative functions and
the pragmatics of language use is still a very new
approach in child language studies. In this approach, language is understood as an information
transmitting system for communicating social
intentions as well as linguistic propositions.
Its eventual goal is an integration of the relationship between the functional and structural
aspects of language acquisition.

A functional analysis begins with the concept

of a speech act or sign act as the basic unit of linguistic communication. Austin (1962) and Searle (1969) state that a speech act is composed of: (1) a linguistic proposition which conveys the conceptual content of an utterance, and (2) an illocutionary force which indicates whether the utterance is an assertion, promise, request, question or other kind of act. The primary determinant of the illocutionary force of an utterance is the communicative intention motivating it. Communicative intentions are expressed through illocutionary acts which convey how a child or adult intends an utterance to be taken. Illocutionary acts are understood primarily in relation to the context and response of the recipient of the communication.

For example, the proposition, "Roger tickle Booee" can be expressed as a request, "tickle Booee, Roger" which communicates the intention that the receiver recognize that this is a solicitation for the performance of an action. The same proposition can also be expressed as other acts, such as a question, "is Roger tickling Booee?" or a description of an event "Roger is tickling Booee" with differing communicative intentions.

Children's ability to communicate intentions appears to be far in advance of their linguistic ability. Dore (1975) and Halliday (1975) have shown that the expression of communicative intentions can be accomplished by very young children, between 6 and 18 months of age, before they develop a lexico-structural system. Dore systematically sampled the spontaneous conversation of children in relatively unrestrained natural settings and classified their utterances according to a variety of illocutionary acts. He found that very young children using only single word utterances produce "primitive speech acts" which function to convey their intentions before they acquire sentences. Bruner (1974) has studied the relationship of prespeech communication to more advanced linguistic forms and has argued that aspects of the communication situation are developmental precursors to the mastery of linguistic structure. Dore (1975) reports that by the time children use sentences their illocutionary acts include requests, questions, agreements and disagreements, statements, descriptions, qualifications, and other acts.

The categories of communicative intentions which Dore has devised can also be applied to the use of sign language by children and apes. An analysis of the communicative intentions expressed by Ally during conversations suggests that his use of signs to communicate his intentions is similar to the speech acts of children in Stage I (Miles 1976). The communicative intentions expressed by Ally included requests for action, requests for attention, questions, descriptions of events or the properties of objects, naming, and statements of feelings or sensations. The classification of Ally's communicative intentions were based on several criteria: (1) the sign communication and its semantic interpretation, (2) the behavioral accompaniments including facial expressions, attention, etc., (3) the social context and activity, and (4) the response of his conversational partner.

Dore (1975) found that action requests and naming were the two largest categories of speech acts in children 3 years of age or younger. This was also true for Booee and Ally, although almost one-fourth of Ally's communications consisted of other kinds of sign acts. The high percentage of action requests and naming in the communicative intentions of both children and apes probably only partly reflects their motivations since Brown (1973) has argued that these communications are often in response to repeated attempts by adults to engage them in conversation exchanges by asking, "what do you want?" or "what's that?". Although sign act information is not yet available for deaf children, it seems likely that they would express similar communicative intentions during Stage I.

Discussion

Thus, on measures of sign performance (form), sign order (structure), semantic relations (meaning), sign acts (function) and sign acquisition (development), apes appear to be very similar to 2 to 3 year old human children learning sign. Except for variations in sign language structure and the age of acquisition of first signs, apes also appear to be very similar to 2 to 3 year old human children learning to speak. In fact, if presented with an unidentified corpus of Stage I utterances, it would not be possible to determine if they were produced by apes or children using current linguistic measures. The only difference is that we know that human children go on to acquire more complicated linguistic operations and we are as yet unsure of the limits of ape linguistic abilities.

These similarities between apes and children support the assertion by Rumbaugh and Gill (1976) that apes and humans share some cognitive abilities underlying the early stages of language acquisition. Although we now know that apes will acquire linguistic skills using a variety of methods, long term acquisition studies (R. Gardner and B. Gardner 1975) suggest that a more human-like social and communicative context enhances the communicative similarities between apes and humans.

Stage II language acquisition in human children involves the modulations of meaning through the control of some grammatical morphemes. The Gardners (B. Gardner and R. Gardner 1975) have attempted to demonstrate evidence for sentence constituents in some of Washoe's sign responses to questions such as "who?" and "where?" which would place Washoe's responses in Stage III. Part of the task of future research will be to explore specific linguistic operations associated with stages of language acquisition to see where apes will diverge from or be exceeded by human children.

It would also be useful to determine whether apes process their linguistic and cognitive skills in ways that are similar to humans. The issue of the nature of homologous behaviors is complicated, but it would be useful to know if apes are using the same brain processes as humans in carrying out their linguistic operations, since the use of similar processes would reinforce the notion that ape and human linguistic and cognitive abilities are closely linked by common origins. The communicative aspects of ape language skills in the laboratory are also a potential source for hypotheses concerning the evolution of language in hominids. For example, the well developed skills of apes in expressing communicative intentions, productive semanticity and environmental reference suggests that these communicative skills could have formed a base which was later combined with new cognitive organizing principles, coming perhaps from greater social complexity (Tanner and Zihlman 1976) or tool production (Lieberman 1976).

Through a new emphasis on communication and the social context of language development, investigators have discovered what mothers have always known — that even very young children are communicating a great deal of information through very simple constructions. Examining the social context within which ape sign language communications with humans and each other take place may help identify important features of ape natural communications. Locating the linguistic and communicative skills of apes within their social context may also lead to a better understanding of the inter-relationships of language, intelligence, communication and cognition. Levin and McShane (1976) have made the important observation that it may be more important to know what apes are doing with their new linguistic code than it is to analyze their performance for the presence or absence of human-like linguistic features. The fact that apes initiate linguistic interactions, "talk" to themselves, ask for what they want and comment on their environment may prove to be more significant than the details of their encoded communications. Menzel (1976) has suggested that the most significant feature of communication research with apes may be to diminish the "racist anthropocentric" view we have concerning our own uniqueness. Of course, there are species differences, linguistic and otherwise, between apes and humans. But the time has passed when we can uncritically apply a double standard for ape and human behaviors. What is called "the method of rich interpretation" with humans is labeled as "anthropomorphizing" with apes. Why should more conservative theoretical and behavioral criteria be applied to apes and not also to human children? The enduring impact of language research with apes may be a shift in our worldview in which we come to realize that we share our world, and perhaps our universe, with other intelligent thinking and communicating creatures.

References Cited

Austin, J.
    1962 How to do Things with Words, New York:
          Oxford University Press.

Bellugi, U. and E. W. Klima
    1976 "Two Faces of Sign: Iconic and Abstract"
          in S. R. Harnad, H. D. Steklis, and J.
          Lancaster (eds.), Origins of Evolution
          of Language and Speech: Annals of the
          New York Academy of Science 280.514-38.

Bloom, L.
    1970 Language Development: Form and Function
          in Emerging Grammars, Cambridge: M.I.T.
          Press.
    1974 One Word at a Time: The Use of Single
          Word Utterances Before Syntax, The Hague:
          Mouton.

Bowerman, Melissa
    1973 Early Syntactic Development: A Cross
          Linguistic Study with Special Reference
          to Finnish, London: Cambridge University
          Press.

Braine, M.
    1963 "The Ontogeny of English Phrase Struc-
          ture: The First Phase" Language 39.1-13.

Brown, Roger
    1973 A First Language: The Early Stages,
          Cambridge: Harvard University Press.

Brown, R. and U. Bellugi
    1964 "Three Processes in the Acquisition of
          Syntax" Harvard Educational Review 34.
          133-51.

Brown, R. and C. Fraser
    1963 "The Acquisition of Syntax" in Charles
          N. Coger and Barbara Musgrave (eds.),
          Verbal Behavior and Learning: Problems
          and Processes, New York: McGraw-Hill.

Bruner, Jerome
    1974 "The Ontogenesis of Speech Acts" Journal
          of Child Language 2.1-19.

Campbell, R. and R. Wales
　　1970　"The Study of Language Acquisition" in
　　　　　J. Lyons (ed.), New Horizons in Linguis-
　　　　　tics, Harmondsworth: Penguin.

Chafe, W.
　　1970　Meaning and the Structure of Language,
　　　　　Chicago: The University of Chicago Press.

Clark, Eve
　　1973　"What's in a Word? On the Child's Ac-
　　　　　quisition of Semantics in His First
　　　　　Language" in T. E. Moore (ed.), Cognitive
　　　　　Development and the Acqsution of Lan-
　　　　　guage, New York: Academic Press.

Dore, John
　　1975　"Holophrases, Speech Acts, and Language
　　　　　Universals" Journal of Child Language
　　　　　2.21-40.

Fillmore, C.
　　1968　"The Case for Case" in E. Bach and R. T.
　　　　　Harms (eds.), Universals in Linguistic
　　　　　Theory, New York: Holt, Rinehart, and
　　　　　Winston.

Fischer, S.
　　1973　"Two Processess of Reduplication in the
　　　　　American Sign Language" Foundations of
　　　　　Language 9.469-80.

Fouts, Roger
　　1973　"Acquisition and Testing of Gestural
　　　　　Signs in Four Young Chimpanzees" Science
　　　　　180.978-80.
　　1974　"Language: Origins, Definitions, and
　　　　　Chimpanzees" Journal of Human Evolution
　　　　　3.475-82.

Gardner, B. and R. Allen Gardner
　　1971　"Two-way Communication with an Infant
　　　　　Chimpanzee" in A. M. Schrier and F.
　　　　　Stollnitz (eds.), Behavior of Nonhuman
　　　　　Primates, Vol. 4, New York: Academic
　　　　　Press.
　　1974　"Comparing the Early Utterances of Child
　　　　　and Chimpanzee" in A. Pick (ed.), Min-
　　　　　nesota Symposium on Child Psychology,
　　　　　Vol. 8. Minneapolis: University of Min-
　　　　　nesota Press.

1975   "Evidence for Sentence Constituents in the Early Utterances of Child and Chimpanzee" Science   104.244-67.

Gardner, R. Allen and B. Gardner
1974   "Review of R. Brown's A First Language: The Early Stages" American Journal of Psychology 87.729-36.
1975   "Early Signs of Language in Child and Chimpanzee" Science   187.752-3.

Halliday, M.
1975   Learning How to Mean, London: Edward Arnold.

Hoffmeister, Robert et al.
n.d.   "Some Procedural Guidelines for the Study of the Acquisition of Sign Language Studies."

Hymes, D.
1971   "Competence and Performance in Linguistic Theory" in R. Huxley and E. Ingram (eds.), Language Acquisition: Models and Methods, London: Academic Press.

Klima, E. S. and U. Bellugi
1970   "Report" in G. A. Miller (ed.), Forum Series on Psychology and Communication, Washington, D. C.: Voice of America.
1972   "The Signs of Language in Child and Chimpanzee" in T. Alloway, L. Krames, and P. Pliner (eds.), Communication and Affect: A Comparative Approach, New York: Academic Press.

Lieberman, Philip
1986   On the Origins of Language, New York: Macmillan.

Lieven, Elena and John McShane
1976   "Language is a Developing Social Skill" paper presented at the Sixth Congress of the International Primatological Society, Cambridge, England, August 23-27.

Menzel, Emil W.
1976   "Implications of Chimpanzee Language — Training Experiments for Primate Field Work — and Vice Versa" paper presented

at the Sixth Congress of the International Primatological Society, Cambridge, England, August 23-27.

Miles, Lyn W.
    1976   "The Communicative Competence of Child and Chimpanzee" in S. R. Harnad, H. D. Steklis, and J. Lancaster (eds.), Origins of Evolution of Language and Speech: Annals of the New York Academy of Science 280.592-7.
    1977   "Sign Language Conversation with Two Chimpanzees" unpublished doctoral dissertation, University of Connecticut, Storrs, Connecticut.
Miller, W. and S. Ervin
    1964   "The Development of Grammar in Child Language" in Monographs of the Society for Research in Child Development 29.9-34.

Mindel, E. D. and M. Vernon
    1971   They Grow in Silence: The Deaf Child and His Family, Silver Spring: National Association for the Deaf.

Patterson, Francine
    n.d.   "The Gestures of a Gorilla: Language Acquisition in Another Pongid Species" unpublished manuscript.

Premack, David
    1971   "On the Assessment of Language Competence in the Chimpanzee" in A. M. and F. Stollnitz (eds.), Behavior of Nonhuman Primates, Vol. 4, New York: Academic Press.

Rumbaugh, Duane
    1973   "Reading and Sentence Completion by a Chimpanzee (Pan)" Science 182.731-3.

Rumbaugh, Duane and Timothy Gill
    1976   "The Mastery of Language-Type Skills by the Chimpanzee (Pan)" in S. R. Harnad, H. D. Steklis, and J. Lancaster (eds.), Origins of Evolution of Language and Speech: Annals of the New York Academy of Science 280.562-78.

Ryan, J.
    1974   "Early Language Development" in M. P. M.

Richards (ed.), The Integration of a Child into a Social World, Cambridge: Cambridge University Press.

Schlesinger, I. M.
1971 "Production of Utterances and Language Acquisition" in D. Slobin (ed.), Ontogenesis of Grammar, New York: Academic Press.

Schlesinger, H. S. and K. P. Meadow
1971 Deafness and Mental Health: A Developmental Approach, Langley Porter Neuropsychiatric Institute.
1973 Sound and Sign, Berkeley: University of California Press.

Searle, John
1969 Speech Acts, Cambridge: Cambridge University Press.

Stokoe, William C.
1976 "Sign Language Autonomy" in S. R. Harnad, H. D. Steklis, and J. Lancaster (eds.), Origins of Evolution of Language and Speech: Annals of the New York Academy of Science 280.505-13.

Tanner, Nancy and Adrienne Zihlman
1976 "The Evolution of Human Communication: What Can Primates Tell Us?" in S. R. Harnad, H. D. Steklis, and J. Lancaster (eds.), Origins of Evolution of Language and Speech: Annals of the New York Academy of Science 280.467-80.

5

Sign Language in Chimpanzees
Implications of the Visual Mode and the Comparative Approach

Roger S. Fouts

The title of this symposium, An Account of the
Visual Mode: Man vs. Ape, has two implicit notions
which I will initially address in this paper. The
first part, An Account of the Visual Mode, tacitly
implies a comparison to the auditory mode. The second, Man vs. Ape, on the other hand implies a comparison that would emphasize the differences between the two species.

With the permission of the Organizer, Fred C.
C. Peng, I will reverse the order and address the
notion that differences between species should be
emphasized first; namely, Man vs. Ape. This in itself is a worthy endeavor. Studying the differences between species is as integral a part of any
comparative science as is studying the similarities.
Since a comparative science often makes inter-
specific comparisons, the differences should be obvious — why else would the organisms be classified
as different species, if they were not different
from one another in some important characteristics.
Similarities, however, are just as important in de-
fining a species. Speaking in this general sense,
in regard to classification of species, might give
one the impression that scientists who compare
species are blind to prejudice in their comparisons
and take into account both the similarities and
differences. However, this has not always been the
case, when Homo sapien happens to be one of the
species in the comparison. Similarities are generally noted in regard to morphological characteristics — the body. But it has been quite a different story where behavior — the product of the
mind — is concerned. In other words, we do not

find it frightening or disconcerting, when a scientist discovers that we are not unique in regard to the possession of hair, eyes, or fingers; nor are we particularly upset, when we discover that the presence or absence of a morphological characteristic we previously thought to be unique to man is shared by another animal (such as the absence of the os penis in the woolly monkey). But when the similarities concern behavior, which has been philosophically viewed as a product of our rational soul or mind, then, individuals tend to become quite defensive about our uniqueness (Mortimer J. Adlers' 1975 article entitled "The Confusion of the Animalists" is an excellent example of this).

It is my opinion that this defensiveness has its historical roots in philosophy, beginning with the dichotomy proposed by Plato that only man has a rational soul, a notion further supported by Descartes' dichotomous approach to man vs. animals. This approach to nature is presently evinced by philosophers, such as Mortimer J. Adler, and linguists, such as Noam Chomsky, among others. This position is at odds with present-day evolutionary thought, because it implies that a continuity exists between species only for morphological characteristics and that evolutionary continuity does not apply to human behavior which is different in kind from that of other species. This position is not surprising, when one considers that Charles Darwin's book The Origin of Species (1859) was only published a little over 100 years ago, and one cannot expect rapid change in what has been the prevailing mode of thought for over 2,000 years.

My own research has been criticized for dealing with analogous or homologous behavior in humans and apes and for ignoring the differences between them. I personally think that emphasizing the similarities and differences are important; and, by my own interest in similarities I do not mean to imply that differences do not exist. Obviously, there are important differences between apes and humans; otherwise, they would not be assigned to different species and genus classifications. It is a matter of scientific interest on my part, however, to study the similarities at a behavioral level, although, historically, the reverse approach has for the most part been adopted. A great deal of literature dealing with this body of research has used the absence of evidence to determine

differences. For example, Noam Chomsky (1967 &
1968) feels that language is the result of a uniquely innate structure in man's brain and that language
is beyond the otherwise intelligent ape. Eric
Lenneberg (1967 & 1971) relies on the lack of physiological or neurophysiological evidence in the
brain of nonhuman primates in order to support a
uniqueness notion. Dobzhansky (1972) stresses
man's uniqueness in terms of psychological properties rather than morphological or physiological
properties. He agrees with Simpson (1964) in regard to the crucial psychological property. "Language is also the most diagnostic single trait of
man: all normal men have language; no other now
living organisms do" (1972:419). Bronowski and
Bellugi (1970) list five characteristics of language. They state that the chimpanzee Washoe has
met the first four but not the fifth — reconstitution. They define reconstitution as being made
up of two procedures: first, analysis of messages;
and, second, synthesis, in which parts of messages
are rearranged to form other messages. Next, they
state that "What the example of Washoe shows in a
profound way is that it is the process of total reconstitution which is the evolutionary hallmark of
the human mind. And for which so far we have no
evidence in the mind of the nonhuman primate even
when he is given the vocabulary ready made" (1970:
673). The list of such examples could go on and on.
Some scientists have admitted their blunders and
others continue to cling to their pre-empirical
data opinions. These types of statements and positions bring to mind the quote of J. L. Austin that
J. J. Gibson (1966) used to introduce his book entitled <u>The Senses Considered as Perceptual Systems</u>:
"There is nothing so plain boring as the constant
repetition of assertations that are not true." However, I would also like to emphasize that there
have been and are now many scientists who also find
the above approaches erroneous.

It is my opinion that the differences between
man and other animals have been over-emphasized
and that this has occurred traditionally within a
Cartesian framework. One of the more noteable exceptions to this approach has been the work of
Jane van Lawick-Goodall with wild chimpanzees. Recently, van Lawick-Goodall (1975) has examined some
potential precursors of behavioral characteristics
that have been considered unique to man. She

examined such characteristics as constant sexual
receptivity, upright posture, altruism, religion,
death, and language. She does not come to any definite conclusions, but in a general sense it appears that precursors to these behaviors in man may
very well exist in our closest living relative,
the chimpanzee.
Along with the behavioral evidence that chimpanzee behavior is much closer to human behavior
than previously thought, there is biochemical evidence which demonstrates extreme similarities in
the blood protein, amino acids, and blood immuno-
logy of the two species. King and Wilson's (1975)
study has shown on the basis of the above mentioned
variables that the chimpanzee is about 0.2% different from humans in terms of these biochemical
measures on the blood of the two species. In contrast, the gorilla is about 0.8% different from
humans on these same measures.

In any event, the historical tradition of emphasizing only the differences between humans and
apes seems to be coming to an end as a result of
laboratory and field studies which have proved that
many of the assumptions concerning differences were
erroneously based on the absence of evidence or negative evidence. Perhaps, now, we can begin to
make up for the dearth of scientific information
caused by the philosophical myopia of the past in
respect to the behavioral similarities between
humans and other species.

Now, I will address the emphasis of the visual
mode in the title of this symposium. This tacitly
implies that it is being compared to the auditory
mode, especially when dealing with communication or
language. And when the modes of language are being
considered, the auditory mode has been examined so
extensively that individuals have often confused
language with speech and vice versa. This has probably resulted from the fact that most of the
scientists who have studied language have only
studied this behavior in one mode and, to make matters worse, they have only studied it in one species. The present situation of the linguists is
analogous to the situation of the experimental psychologists of twenty years ago. But instead of ad-
dressing the relatively broad area of learning, the
linguists have limited themselves to a narrower
area of behavior; namely, oral language. And in

place of the white rat and a few other species, the
linguists study only one species — Homo sapien.
The naive behaviorism of psychology's past led to
the rather myopic view that all behavior was controlled by the environment through learning. This
position resulted from the fact that psychologists
stuidied very few behaviors other than learned ones.
So, experimental psychologists began to assume that
everything was learned, since they did not bother
to examine other behaviors, such as maternal behavior, reproductive behavior, feeding behavior and
so on.

Some linguists have accomplished a similar
feat in regard to language. By studying only one
species, and in fact often only the normal adult
members of one species, they have frequently come
to the conclusion that the particular behavior they
were studying existed only in that particular
species. The dangers of this intellectually crip-
pling approach of "behavioral apartheid" should be
obvious. In this vein, it is interesting to find
that some linguists view comparative studies of
language in other species with disdain rather than
welcoming the input from an area that they have
ignored. It is not surprising, then, to find that
of the six research projects around the country
examining language behavior in apes all were
started by psychologists and not linguists.

This same prejudice has been revealed in the
emphasis on the auditory mode in linguistic research and in the implicit assumption that this is
the only mode in which language can be expressed.
This is often evidenced in the ignorance of the
visual mode of communication used by the deaf, i.e.,
sign language which, since it has not been studied
extensively in the linguistic literature, many
people have assumed is not a language.

In regard to sign language and linguistics, a
little knowledge can sometimes be dangerous. For
example, Mounin (1976), who studied American Indian
sign language and found that it did not have duality of patterning, assumed that the same was true
for American Sign Language for the deaf. This
makes as much sense as saying that English does not
have the copula because Hebrew does not have it.

The above example once again illustrates the
egocentric myopia that results in the ignorance of

a linguistic behavior, if it embraces a modality
other than the auditory one. It is almost as if
language is viewed as being above the laws of learning that apply to other behaviors and that the auditory mode is an integral part of language rather
than simply one mode of expression for a rather complex cognitive behavior. In this context, Harvey
Sarles has stated the following: "Biologically
speaking, we expect continuity and relationship,
not emergence and saltation and we are rightfully
suspicious when told that man is more than a bit
outside of nature" (1972:4).

It is my view that the particular mode of expression for language behavior is not as important
as the underlying cognitive capacities that allow
for the production of language behavior. As Hornbostel stated about the senses: "It matters little
through which sense I realize that in the dark I
have blundered into a pigsty" (1927:83). The same
is true of language and communication. For example,
in regard to animal behavior, Peter Marler (1965)
makes the following statement: "In most situations
it is not a single signal that passes from one animal to another but a whole complex of them, visual,
auditory, tactile, and sometimes olfactory. There
can be little doubt that the structure of individual
signals is very much affected by this incorporation
in a whole matrix of other signals." I would maintain that the particular sense mode I use to communicate or the particular sense mode through which
I understand a message matters little in regard to
the intention of the message or the underlying cognitive capacities which allow me to produce and receive the message.

Recent experimentation also points out that
the comprehension of phonetic boundaries in human
English speaking adults may not be something unique
to humans. Kuhl and Miller (1975) have found that
chinchillas were able to make a voiced-voiceless
distinction in the alveolar plosive consonants of
[d] and [t]. And, it is commonly known that parrots
and other birds are able to produce excellent imi-
tations of the human voice. It is probable, however, that these nonhuman animals are using areas
of their brain that are not homologous to the areas
of the brain that humans use for vocalizing. Chimpanzees, on the other hand, can comprehend vocal
English. In a pretest of a study examining the
use of vocal English words to teach signs, Fouts,

Chown, and Goodin (1976) found that their chimpanzee subject was able to comprehend the ten English words used in their study. Other researchers have noted the ability of the chimpanzee to comprehend words and phrases in English (Kellogg and Kellogg 1933; Hayes 1951).)

In regard to primates in general, it can be said that the arboreal species rely primarily on auditory communication, perhaps because visual communication is inefficient in the dense foliage of the tree tops. But the more terrestrial primates often rely on visual communication more than the arboreal species, perhaps because there are less visual obstructions to occlude their signals. In regard to primate vocalization, I think it can be safely said that they apparently do not have as much plasticity in their vocalization as man. Why does terrestrial man rely on an auditory mode? Was it because he was such a small creature that he spent most of his time hiding in tall grasses and bushes? These speculations are amusing in the sense of parlor games. But this type of speculation is based on the assumption given to us by linguists that man's primary mode of communication is in the auditory channel. Perhaps, this assumption is incorrect, and as I stated earlier, more a result of the linguists' intellectual myopia than of empirical fact. Had linguists looked more carefully at their chosen species, they might well have come to a different conclusion. For example, the research concerning nonverbal communication comes to a quite different conclusion. If the estimates of Ray Birdwhistell (1955) and Albert Mehrabian (1968) are true, nonverbal communication, which is largely visual with some audition, is responsible for 55 to 75% of the meaning generated in a two-person conversation. Sarles (1972) thinks that it may come to as much as 80%. This once again points to the dangers involved in looking at only one narrow type of behavior and ignoring the total process. (It may be that the visual mode is extremely important as an integral part of man's communication systems.)

Charles Darwin (1872) in his book entitled The Expression of the Emotion in Man and the Animals viewed communication behavior as being continuous in much the same fashion as morphological characteristics. In other words, a continuity of mind

exists as well as a continuity of body. The researchers examining nonverbal communication and others,
such as J. A. R. A. M. van Hooff (1972), have taken
Darwin seriously: van Hooff, for example, examines
the phylogeny of laughter and smiling. But these
people are few compared to scientists who do not
seem to accept the notion of continuity in behavior
as well as morphology. Perhaps, there is hope for
improvement in the next generation.

Now, I will briefly examine some of the recent
research we have completed at the Institute for
Primate Studies at the University of Oklahoma and
other related researches. This is but a small sample of the researches going on around the country.
David Premack (1971) has used various shaped and
colored pieces of plastic to represent words in
order to study the negative article, the interroga-
tive, Wh-questions, the concept of "name of," dimensional classes, prepositions, hierarchically organized sentences and the conditional. Rumbaugh,
Gill, and von Glasserfeld (1973), using computer
keys of various colors with lexigrams on the keys,
have demonstrated that their chimpanzee is able to
produce sentences, to complete grammatically correct imcomplete sentences and to reject those in-
complete sentences that are not grammatically correct. Gill and Rumbaugh (1974) have also examined
naming skills in their chimpanzee and found that
their subject has learned that things can be referred to by name. Other research has used an
existing human language, American Sign Language for
the Deaf (ASL), in order to establish a two-way communication with nonhuman primates. Allen and
Beatrice Gardner (1971) were the first to do this
with their chimpanzee Washoe. They found that Washoe was able to acquire the signs and use them in
contextually correct situations. They also found
evidence suggesting that Washoe's early combinations
had the rudiments of syntax. They studied Washoe's
use of Wh-questions, her vocabulary reliability,
compared her early utterances to those of children
(Gardner and Gardner 1974), and compared sentence
constituents in her early utterances to those of
children (Gardner and Gardner 1975a). The Gardners
(1975b) have recently begun a second study using
several young chimpanzees. Terrace and Bever (1976)
are also beginning a project using sign language
with a chimpanzee. They hope to explore the expression of internal states using sign language.
And, of course, Penny Patterson, who has a paper

in this volume, is using sign langauge with a gorilla. So, it can be seen that since the early beginning of the Gardners, ten years ago, the researches
in this area have grown extensively.

The Fouts, Chown, and Goodin (1976) study mentioned earlier in this paper is particularly relevant to this symposium, since it concerns both the
auditory and visual modes. Several of the chimpanzees we work with have been reared in human homes
and as a result they have an understanding of vocal
English. In this study, we examined the relationship between the chimpanzee's understanding of vocal English and his production of ASL. Initially,
the subject, Ally, was tested on his ability to
understand ten vocal English commands (e.g., "Pick
up the spoon"). After Ally met the criterion of
correctly following five vocal commands for each of
the ten objects, he was then taught a sign for vocal
English word in a blind condition to control for
cueing. Next, he was tested to see if he could
transfer the signs, taught to him using a vocal
English word, to the physical object representing
them. Ally was able to do this for all ten words.
At one level, this is comparable to second langauge
acquisition in humans. At another level, it has
cross-modal implications, if the study is viewed in
the following manner. Ally initially assocaited a
visual object with an auditory stimulus; next, he
acquired a gestural response to the auditory stimulus and, then, finally in testing he transferred
this gestural response to a visual stimulus. This
is evidence that a chimpanzee can comprehend vocal
English and transfer a response from the auditory
modality to the visual modality.

The implications for this symposium are that
the chimpanzee is capable of comprehending vocal
English. Thus, the main difference between chimpanzees and humans on this topic is not the ability
to comprehend vocal English, but rather the chimpanzees' relative inability to produce vocal
English. The reasons for this difference could
range from peripheral differences in the structure
of the oral cavity and the lack of tongue mobility,
as was noted by Lieberman, Crelin, and Klatt (1972)
and is reviewed and elaborated by Peng in this volume, to possible differences in the brain, as was
noted by Lenneberg (1967). But the ability of
chimpanzees to comprehend vocal English suggests
that they have the cognitive capacity necessary for

understanding vocal language and this would also
suggest that they have the capacity to produce a
language, if the visual mode is used. The main
difference seems to lie in the ability of a chimpanzee to use the auditory mode as a vehicle for
the complex cognitive behavior of langauge production. As I stated earlier, the apes do not seem to
have the extreme degree of plasticity in terms of
producing vocal behavior as do humans. Most chimpanzee vocalizations seem to be emotive in nature
and are often elicited by stimuli in the environment. However, I have observed a chiampanzee,
Booee, using a vocalization (largely voiceless) in
place of a sign. This vocalization would be analogous to laughter in humans. For example, instead
of signing YOU TICKLE BOOEE, Booee would often sign
YOU and, then, vocalize with the laughing sound and,
finally, sign BOOEE. There is obviously some plasticity in the chimpanzees' vocal ability, but certainly not to the same degree as in humans. Hayes
and Nissen (1971) taught their chimpanzee, Viki,
to speak four vocal English words, after working
with her for six years. This evidence would support the preceding point.

Given the chimpanzees' apparent problems with
vocal production, the next logical step is to see
if they can produce language in another modality.
The Gardners and others have attacked this problem.
But once this is done, we immediately run into the
problem of not knowing what language is, in definit-
ional terms. At times, there seems to be almost as
many definitions of language as there are linguists,
though this is of course an exaggeration. Since
there is no generally accepted definition of language, I will posit my own opinion about the nature
of language. I personally think that language is
just another behavior subject to the same laws as
are other behaviors. However, it does seem to be
rather complex. My definition of language is extremely similar to the characteristic of language
that Bronowski and Bellugi (1971) referred to as
reconstitution which is " ... a procedure of analysis, by which messages are not treated as inviolate
wholes but are broken down into smaller parts, and
a procedure of synthesis, by which the parts are
rearranged to form other messages" (1971:670). I
would make one major change in their definition;
that is, to change the word "message" in the passage to "observable behaviors" and add that these
behaviors are used communicatively. The reason for

the word change is that I do not think that this
ability is unique to language but, instead, is common to many behaviors that are not traditionally
classified as a form of communication. In other
words, it is a function of the cognitive abilities
of the organisms. This definition is also similar
to two of Hockett's (1958) key properties of language; namely, openness and productivity.

The Fouts, Chown, Kimball, and Couch (1976)
study is particularly relevant to this definition.
This study was designed to examine the active application of a syntactic system in responding correctly to signed commands and, in a second study,
in responding to the spatial arrangements of objects by signing descriptive sequences of signs in
ASL. In the first experiment, the subject, Ally,
was first taught to obey a corpus of 33 signed commands. Each command requested Ally to select an
object from a box containing five objects and, then,
to deliver it to one of three locations. New commands were, then, formed by vocabulary substitutions at the object and location positions. Ally's
comprehension of the new commands were tested using
a double-blind procedure to control for the possibility of cueing by the humans. The results indicated that Ally could respond correctly to signed
commands he had never seen before. In the second
study, Ally was taught to construct sequences of
signs describing the relations "on," "in," and "under" between objects arranged before him. He was,
then, tested on relationships presented to him
using a double-blind procedure which varied from
completely new relationships (new subject, new location) to completely familiar relationships (training relationships). The results of this experiment
indicated that Ally could master a simple syntactic
system. By correctly describing approximately
equal percentages of test and training relationships under test conditions, it can be inferred
that Ally was in control of a simple productive
grammar. He used this grammar actively and spontaneously in response to objects arranged before
him to exemplify relations he had previously
learned. Taken together, the results from both
experiments demonstrate that a chimpanzee can
evince the rudiments of the linguistic property of
productivity, in both comprehension and active application of the system. Ally was not treating
the commands as inviolate wholes; rather, he was
able to actively produce new sequences of signs by

recombining the various parts to describe new relationships. This demonstrates that a chimpanzee
is capable of acquiring language characterized by
properties at the least very important to our own
language behavior, if not essential to language
itself.

Let me summarize now that I have, in the preceding, attempted to make the following points.
The first was that there are behavioral as well as
morphological similarities between apes and humans
and the behavioral similarities have often been
ignored in favor of emphasizing the differences.
We should consider both of the similarities along
with the differences at a comparative behavioral
level whenever we study any behavior. The second
point I made was that the auditory mode has been
over-emphasized in regard to language. Language
should be viewed as a cognitive ability which trans-
cends any particular modality that might be used to
express it.

Finally, I examined the role of the visual and
auditory modes in language behavior in chimpanzees.
I pointed out that chimpanzees certainly seem to
have the cognitive ability for language but that
they have not yet demonstrated the plasticity in
vocal production necessary for speech. Next, I
described a study which demonstrated that an essential characteristic of language behavior, if not the
essential characteristic of language itself, could
be mastered by a chimpanzee under test conditions.

References Cited

Adler, Mortimer J.
    1975 "The Confusion of the Animalists" in
          Robert M. Hutchins and M. J. Adler (eds.),
          The Great Ideas Today 1975, pp. 72-89,
          Chicago: Encyclopedia Britannica, Inc.

Birdwhistell, R. L.
    1955 "Background to Kinesics" ETC Review of
          General Semantics 13.10-5.

Bronowski, J. and U. Bellugi
    1970 "Language, Name, and Concept" Science
          168.669-73.

Chomsky, N.
    1967 "The Formal Nature of Language" in E.

Lenneberg (ed.), <u>Biological Foundations</u>
<u>of Language</u>, pp. 397-442, New York: John
Wiley.
1968   <u>Language and the Mind</u>, New York: Har-
court, Brace and World.

Darwin, C.
1859   <u>The Origin of Species</u>, London: John
Murray.
1872   <u>The Expression of the Emotions in Man</u>
<u>and the Animals</u>, London: John Murray.

Dobzhansky, T. H.
1972   "On the Evolutionary Uniqueness of Man"
in T. H. Dobzhansky, M. K. Heckt, and
W. C. Steere (eds.), <u>Evolutionary Bio-</u>
<u>logy</u>, pp. 415-30, New York: Appleton-
Century-Crofts.

Fouts, R. S., B. Chown, and L. Goodin
1976   "Transfer of Signed Responses in American Sign Language from Vocal English
Stimuli to Physical Object Stimuli by
a Chimpanzee (Pan)" <u>Learning and Moti-</u>
<u>vation</u> 7.458-75.

Fouts, R. S., W. Chown, G. Kimball, and J. Couch
1976   "Comprehension and Production of Ameri-
can Sign Language by a Chimpanzee (Pan)"
paper presented at the XXI International
Congress of Psychology in Paris, France,
July 18-25.
Gardner, B. T. and R. A. Gardner
1971   "Two-way Communication with an Infant
Chimpanzee" in A. M. Schrier and F.
Stollnitz (eds.), <u>Behavior of Nonhuman</u>
<u>Primates</u>, Vol. 4, New York: Academic
Press.
1974   "Comparing the Early Sentence Utterances
of Child and Chimpanzee" in A. Pick (ed.),
<u>Minnesota Symposium on Child Psychology</u>,
Vol. 8, Minneapolis: University of Min-
nesota Press.
1975a   "Evidence for Sentence Constituents in
the Early Utterances of Child and Chimpanzee" <u>Journal of Experimental Psycho-</u>
<u>logy: General</u>, 104.244-67.
1975b   "Early Signs of Language in Child and
Chimpanzee" <u>Science</u> 187.752-3.

Gibson, J. J.

1966 The Senses Considered as Perceptual Systems, Boston: Houghton Mifflin Co.

Gill, T. V. and D. Rumbaugh
     1974 "Mastery of Naming Skills by a Chimpanzee" Journal of Human Evolution 3.483-92.

van Lawick-Goodall, J.
     1975 "The Chimpanzee" in V. Goodall (ed.), The Quest for Man, pp. 131-69, London: Phaidon Press limited.

Hayes, C.
     1951 The Ape in Our House, New York: Harper.

Hayes, K. and C. H. Nissen
     1971 "Higher Mental Functions of a Home-raised Chimpanzee" in A. M. Schrier and F. Stollnitz (eds.), Behavior of Nonhuman Primates, Vol. 4, pp. 59-115, New York: Academic Press.

Hockett, C. F.
     1958 A Course in Modern Linguistics, Toronto: MacMillan.

van Hooff, J. A. R. A. M.
     1972 "A Comparative Approach to the Phylogeny of Laughter and Smiling" in R. Hinde (ed.), Nonverbal Communication, pp. 209-37, Cambridge: Cambridge University Press.

von Hornbostel, E. M.
     1927 "The Unity of the Senses" Psychology 7.83-9.

Kellogg, W. N. and L. A. Kellogg
     1933 The Ape and the Child, New York: McGraw-Hill.

King, M-C. and A. C. Wilson
     1975 "Evolution at Two Levels in Humans and Chimpanzees" Science 188.107-16.

Kuhl, P. K. and J. D. Miller
     1975 "Speech Perception by the Chinchilla: Voiced-voiceless Distinction in Alveolar Plosive Consonants" Science 190.69-72.

Lenneberg, E. H.
    1967 <u>Biological Foundations of Language</u>, New
          York: John Wiley.
    1971 "Of Language Knowledge, Apes, and Brains"
          <u>Journal of Psycholinguistic Research</u>
          1.1-29.

Lieberman, P., E. Crelin, and K. Klatt
    1972 "Phonetic Ability and Related Anatomy of
          the Newborn and Adult Human, Neanderthal
          Man, and Chimpanzee" <u>American Anthropo-
          logist</u> 74.287-307.

Marler, P.
    1965 "Communication in Monkeys and Apes" in
          I. DeVore (ed.), <u>Primate Behavior: Field
          Studies of Monkeys and Apes</u>, pp. 544-
          84, New York: Holt, Rinehart, and
          Winston.

Mehrabian, A.
    1965 "Communication Without Words" <u>Psychology
          Today</u> 2(4).52-5.

Mounin, G.
    1976 "Language, Communication, Chimpanzees"
          <u>Current Anthropology</u> 17.1-21.

Premack, D.
    1971 "On the Assessment of Language Competence
          in the Chimpanzee" in A. M. Schrier and
          F. Stollnitz (ed.), <u>Behavior of Nonhuman
          Primates</u>, Vol. 4, New York: Academic
          Press.

Rumbaugh, D., T. V. Gill, and E. von Glaserfeld
    1973 "Reading and Sentence Completion by a
          Chimpanzee (Pan)" <u>Science</u> 182.731-3.

Sarles, H. B.
    1972 "The Search for Comparative Variables in
          Human Speech" a symposium paper presented
          at the Animal Behavior Society Meetings,
          Reno, Nevada, June 13-16.

Simpson, G.
    1964 <u>This View of Life</u>, New York: Harcourt,
          Brace and World.

Terrace, H. S. and T. G. Bever
    1976 "What might be learned from Studying

Language in the Chimpanzee? The Importance of Symbolizing Oneself" in S. R. Harnad, H. D. Steklis, and J. Lancaster (eds.), <u>Origins of Evolution of Language and Speech: Annals of the New York Academy of Science</u> 280.579-88.

Language Skills, Cognition, and the Chimpanzee

6

Duane M. Rumbaugh, E. Sue Savage-Rumbaugh,
and Timothy V. Gill

The thesis of this paper is that the evolution of cognition has entailed the selection of
covert psychological processes which served the
interests of social behaviors. This selection has
allowed for increasingly open communication systems
which have led to man's natural languages. But the
covert processes in reference are <u>not</u> unique to
man. Rather, they are presumed to be extant relative to the degree to which a given form of animal
life manifests plasticity (adaptive variations) in
its social behaviors. It is assumed that to the
degree that there is similarity in species' brains
and brain function, there is commonality of cognitive operations. From this it follows that particularly among the great apes (Pongidae) and man,
there are major dimensions of cognition held in
common. It is hypothesized that it is the covert
dimensions of cognition held in common by apes and
man which have allowed for the well-known successful demonstrations of two-way, linguistic-like
communication between apes of language-training
programs and human beings.

<u>Cognition</u> is held by the authors to be an ad-
vanced form of intellectual function. Basically
it provides for the perception of relationships
among the attributes of diverse things and events.
All learning holds the potential for enhancing
adaptation through behavioral alterations. But
whereas classical and instrumental forms of learning result in behavioral alterations through the
selective reinforcement of certain behaviors,
the behaviors so "learned" are essentially basic to
the response repertoire of the organism. Conditioning can and does alter the morphology of re-

*Supported by National Institutes of Health Grants
HD-06016 and RR-00155.

sponses and the occasions for selected responses
to be manifest; however, conditioning does little
more than to re-arrange the basic response elements
of an organism's repertoire and their relative probabilities. By contrast, cognition results in alterations of organisms' behavior patterns because
of the new comprehensions or understandings which
come about through the emergence of perceived relationships, not through the arduous selective reinforcement of certain responses at the expense of
others. Based on generalized experience and a history of discerning similarities and differences
among the attributes of things and events, an organism might become cognitive as some positive
function of the complexity of its brain (Rumbaugh
1970). The emergence of cognition and its atten-
dant functions serve to expand an organism's op-
tions for behaviors which might prove adaptive,
particularly when the challenge is essentially
novel and where old established response patterns
are not likely to be appropriate.

The assertion that life forms other than man
are capable of cognitive processes is not new.
For decades the literature has been replete with
solid evidence for cognition, notably in apes
(Köhler 1925; Yerkes & Yerkes 1929; Rumbaugh
1971 & 1977b; Mason 1976) but also in a wide variety of mammals, including the laboratory rat which
has been known to be capable of reasoning since the
early studies of Tolman (1948) and N. R. F. Maier
(1929). Why is it, then, that there is recent and
intense interest in the question of cognitive processes in the apes? The reasons are probably not
all simple, but one of the primary ones surely is
the fact that the well-known ape-language projects
(Gardner & Gardner 1971; Premack 1971 & 1977;
Fouts 1974; Rumbaugh 1977a) continue to produce
reports of nonvocal subjects doing remarkably well
in linguistic training with language systems that
are manual-visual rather than oral-aural. Because
the apes have done so well in an increasing array
of projects (Patterson n.d.; Terrace & Bever
1976), it is argued that (i) language should not
be viewed as contingent upon speech, (ii) that the
requisites to language are not uniquely human,
(iii) that apes are more "intelligent" than they
have been thought to be historically, (iv) that
the mentality of apes is likely more similar to
man's than has been allowed historically, and (v)

that there is great potential for studying the language-type behaviors of apes and possibly of other animals as well.

The pressures for the evolution of cognition and advanced forms of intelligence were likely numerous, but surely included (i) the need for co-operation, frequently a requisite for success in the procurement of food, security, and shelter, and (ii) the differentiation of roles and "duties" as offspring became increasingly dependent upon pro-longed maternal care for survival and as the asso-ciated span of years for socialization of the off-spring became greater and greater.

Mason et al. (1968; Mason 1970) have specu-lated upon ways in which the social milieu selected for advanced learning skills. The rhesus monkey, that lives in a large troup, comprised perhaps of several hundred, must be highly perceptive and able to identify rapidly the important elements in an ever-changing social scene. Failure to do so most assuredly results in the loss of access to the nourishments essential to life and risk to life and limb from powerful, aggressive adults who in-teract socially as though there is a code of be-havior which must be honored.

Chimpanzee social living is viewed by many researchers as infinitely more complex and de-manding than that of monkeys. How and why is this possible? The answer, we believe, lies in the apes' ability to communicate more complex and ar-bitrary messages — messages which reflect a dif-ferent level of self-other awareness than that ex-perienced by monkeys and which consequently permit the exchange of information regarding intent, in-cluding conditional intent. It is this awareness which thereby frees the ape from the simultaneous experiencing and expressing of emotion. Further-more, it is this same self-other awareness which permits the development of learned representation-al symbolic communicative behaviors.

Such a perspective as this implies that through close study of the chimpanzee's natural gestural communication system, much insight can be gained as to what these skills might be, how they develop, and what they imply about the animal's cognitive understanding of the world around it.

Although there have been several accounts of chimpanzees gesturing to human beings in an elaborate manner (Gardner and Gardner 1971; Hayes and Hayes 1955; Kellogg 1968), the reports of interanimal gesturing indicate that gesturing is not the most common means of communication, and when it occurs, it is typically limited to begging, taking, and embracing motions (van Lawick-Goodall 1968; McGinnis 1972). Initial observations of a recently captured group of pygmy chimpanzees revealed however, that this species regularly gestured to one another. Because such frequent and complex inter-animal gesturing had not been reported for either wild or captive groups of common chimpanzees because it occurred regularly in these pygmy chimpanzees, and because it was often employed to transmit information regarding relatively specific behaviors (copulatory position in particular), gesturing in this species was viewed as a potentially fruitful area of study and one which might have implications for the question of language prerequisites.

The subjects of study were three wild-caught pygmy chimpanzees (Pan paniscus) housed at the Yerkes Regional Primate Research Center and on lend-lease from the Zairian government. At the initiation of the study, these animals had been in captivity for six months.

The pygmy chimpanzee is a little known species of ape which is roughly one-half to two-thirds the size of the common chimps. Pygmy chimpanzees possess greater cranial capacity and a flattened cranio-facial structure. Anatomists familiar with their morphology suggest that they are the best living model for reconstruction of early hominid life.

The first analyses of gestural data from these animals was limited to gestures which preceded copulatory bouts, so that some nonarbitrary means of assessing the intent of the gestures could be employed. Nearly all gestures which preceded copulatory bouts appeared to be indicators of copulatory intent and desired position.

Pygmy chimpanzees, in contrast to common chimpanzees, employ a wide variety of copulatory positions; and prior to each copulatory bout, some

mutual agreement as to position must be reached if
a completed copulatory bout is to occur (Savage and
Bakeman 1976). By looking only at gestures which
preceded copulatory bouts and the positional outcomes of each bout, it was possible to determine
whether particular gestures were more likely to be
followed by one type of copulatory position and
other gestures by another. Without such a method
of assessing gestural outcomes, even though the intended meaning of gestures appears to be clear to
the observer, there is no way to determine whether
or not the chimpanzees interpret the gesture in a
similar fashion, or even if they place any significance at all in the gestures.

A total of 21 gestures were specified and described. Analyses were based on 343 observed copulatory bouts that contained at least one gesture,
for a total of 891 gestures. Live observation
hours totalled 600.

Although the term "gesture" has been used by
others to indicate any specific, stylized body or
limb movement (McGinnis 1972; van Lawick-Goodall
1968), the use of the term here is restricted to
hand and/or upper forelimb motions. While whole
body postures like "hunched shoulder" are communicative, such posturings occur throughout the mamal-
ian order, and do not appear to be specifically
prelinguistic forms of communication as do hand
gestures (Hewes 1977).

Following the identification of 21 individual
gestures and collection of data, gestures were
grouped into one of three broad categories: (1)
positioning motions, (2) touch plus iconic hand
motions, and (3) iconic hand motions. These categories correspond with the presumed evolutionary
sequence of the appearance of such gestures. Cf.
(Table 1).

Positioning movements are the most primitive
or basic type of hand movement and others appear to
be derived from them. They include any use of the
hands to move the recipient's body or limbs in a
direction which will promote the assumption of a
particular posture. During a positioning motion,
the initiator usually held the recipient's limb and
gently pushed it in the desired direction. The actual movement of the limb was done by the recipient

TABLE I

Gestures observed preceding and during
socio-sexual bouts in <u>Pan paniscus</u> grouped into types

1st ORDER GESTURES Positioning Movements	2nd ORDER GESTURES Touch and Iconic Hand and Arm Movements	3rd ORDER GESTURES Iconic Hand and Arm Movements
1. Push limb across body (used to induce partner to turn around)*	1. Touch outside of partner's shoulder, hip or thigh, and motion across body with hand and forearm movement	1. Move hand and forearm across body
2. Push leg or arm out from body; generally performed "en face" (used to move limbs out from partner's venter to facilitate ventro-ventral positioning)*	2. Touch hand or arm and motion outward from partner's body	2. Stand bipedally and wave arms out from body
3. Pull toward self by putting arm around partner's back (used to move partner into proper ventro-ventral position)*		3. Raise arm with palm down
4. Position partner's lower body with both hands (used to induce partner to assume a stance compatible with	4. Rest knuckles on arm or back, and move arm toward self	

(TABLE I CONTINUED)

1st ORDER GESTURES Positioning Movements	2nd ORDER GESTURES Touch and Iconic Hand and Arm Movements	3rd ORDER GESTURES Iconic Hand and Arm Movements
ventro-dorsal copulation; employed after genitalia of female have already been oriented toward male for ventro-dorsal position)*		
5. Pull limb toward self (used to induce partner to move closer to initiator)*	5. Touch shoulder or back and move hand toward self	5. Hold hand toward partner
6. Push under chin (used to induce partner to stand bi-pedally prior to ventro-ventral standing copulatory bout, or to lie on their backs, prior to a prone ventro-ventral bout)*	6. Touch head, chin or inside of shoulder and lift hand upward	6. Raise arm and flip hand up-ward at wrist
7. Walk to other end of cage and gaze at partner (used to induce partner to move to another location prior to copulation)*	7. Touch partner and walk to other end of cage	7. Move hand toward another portion of the cage

*Functions, listed in parentheses, are similar across all three gestural types.

and not the initiator. The initiator merely indicated the desired direction of movement by starting the limb in that direction.

Touch plus iconic hand motions seems to be a less direct form of positioning in that the limb or portion of the body to be moved is lightly touched, then the desired direction of movement is indicated by the iconic hand motion which depicts the intended direction. For example, if the initiator wishes the other animal to turn around, this is so indicated by a turning motion of the hand at the wrist.

Completely iconic hand motions form the third category and, as the heading suggests, the initiator simply indicates, via an iconic hand movement, what he would like the recipient to do. These gestures varied from indicating to the recipient to move his whole body to another location in the cage.

These basic gestural forms and particular instances of each are depicted in Figures 1-4.

Effects of Gesturing

Slow motion analysis of videotaped gestural exchanges strongly suggested that these gestures were not employed randomly. There was often a close correspondence in an iconic sense between gestures and the following body movements of the recipient of the gestures.

In accordance with the hypothesis that the iconic gestures have grown from or are an evolutionary extension of the touch gestures, which are, themselves, an extension of positioning movements, the frequencies of each such gestural category were summed.

Positioning gestures, thought to be most primitive and basic communicators of position, were the most frequent. Completely iconic hand motions, the most highly evolved position indicators, were the least frequent; and the touch gestures fell in-between these two, as would be expected. The preponderance of positioning gestures suggests that this form may still be the most effective method of communication among pygmy chimpanzees. Furthermore, they suggest that inter-animal communication

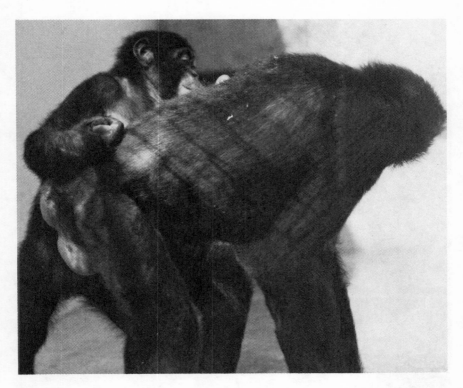

Figure 1. The young male stands bipedally and uses "positioning movements" to position the female for ventro-dorsal copulation. He is attempting to rotate her genitalia with his right hand laid across the lower portion of her back. She has her head turned to him, maintaining eye contact as is typical of pygmy chimpanzees during sexual interactions.

Figure 2. The female does not respond to the gesture in Figure 1., so the male moves in front of her and uses a "touch plus iconic motion" by tapping on her shoulder and gesturing away from himself.

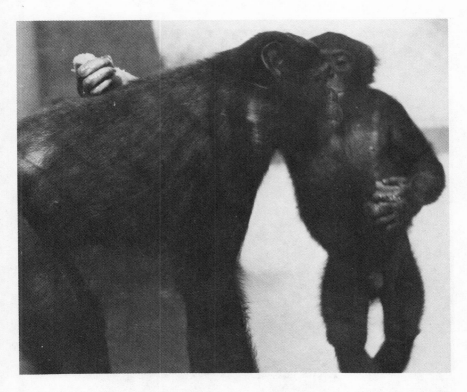

Figure 3.   The female still refuses to present dor-
sally, so the male employs both a "position gesture" (right hand) and an "iconic" gesture to turn
around (left hand).

Figure 4. Finally, the male protrudes food from
his mouth, touches the female on the inside of
arm and gestures outward to encourage her to move
her arms out and present her venter. She did and
ventro-ventral copulation ensued.

about socio-sexual interactions may have been an important evolutionary pressure in the development of complex cognitively based communicatory skills.

Requesting, via an iconic gesture, that another animal move its body through space to a particular location of position is not a simple thing cognitively. It requires: (i) a clear concept of self and others, (ii) the realization that personal desires can be communicated to another individual, (iii) that there is an equivalence between the motion of the hand and the movement of the recipient's body, (iv) that the hand is not acting as a hand in the instance of gesturing, but as a temporary symbol for the recipient's body.

These types of symbolic skills are often cited as important language prerequisites (Bricker and Bricker 1974; Miller and Yoder 1974), and suggest that pygmy chimpanzees may indeed be important living models for pre-language theory and study.

From the perspective of the evolution of cognitive skills, the most intriguing and important aspect of gestural communication in the pygmy chimpanzee is that in their gestural system is displayed what appears to be a most important step in the evolution of linguistic ability — the transition from very concrete expressive forms to the beginnings of arbitrary symbolic gestural expression.

The gestural propensity and capability of the pygmy chimpanzee far exceeds that yet reported for any other ape. This communicatory skill, coupled with the large brain and gracile skeletal structure, suggests that Pan paniscus is, perhaps, the best living animal model to which we may look in the reconstruction of the history of our own species. The frequent and complex gestural skills demonstrated by these animals lend strong support to a gestural origins theory of language evolution as propounded by Hewes (1973 & 1977).

Iconic gestures are of significance in that they are neither incipient acts nor suggestions of that which the signaller is about to do; rather, they serve to indicate that which the signaller would have the recipient do. A "come" gesture, the movement of the hand toward the body (McGinnis

1972) is not an incipient approach movement by the
signaller, but rather a representation of that
which the signaller would have the recipient do in
turn. It suggests an encoding of "intent" on the
part of the signaller into a behavior of apparent
representational value to which another, the recipient, might respond or after which it might pattern its behavior (Krauss 1977; Carroll 1974).
The majority, if not all, of the self-generated
signs of chimpanzees in sign-language studies are
highly iconic, as with the example of Washoe's sign
for bib— fingers drawing the outline of one on the
chest (Gardner & Gardner 1971). But iconic gestures cannot refer to the purely abstract as can
arbitrary signals and words. How is it, then, that
the ape-language projects have succeeded in the development of language skills in apes that, as far
as is known, do not evidence arbitrary symbolic
language skills in their natural habitats?

The answer rests, we believe, in the stated
presumption that apes are pre-adapted by virtue of
the advanced covert cognitive skills which have
evolved to mediate the social demands of their respective species. The ape-language projects avail
to the ape subjects a simplified public language
system in which agreement can be reasonably achieved between man and ape as to the meanings of
words and how their meanings can be modulated
through their ordering into sentences. Only with
the assistance of man can the ape come to conca-
tenate arbitrary signals words into meaningful
phrases and sentences.

Human language systems allow for the ultimate
separation between events and the communications
which are in reference to them. And in the ape-
language projects, it is not at all uncommon for
the apes to "talk" about that which is not present-
evidence of significant ability to separate events
from specific times.

Evidence that the apes of various projects
learn words with increasing facility, that they
will "coin" names for things in ways that suggests
their sensitivity to the salient characteristics
of that named, that they have an ability to master
the rudiments of syntax, and that they have a read-
iness to extend their language skills quite appropriately beyond the specific contexts within which

those skills were learned has been reviewed else-
where (Rumbaugh 1976 & 1977a). These and other
abilities of a verbal nature are believed to rest
on the same cognitive base as do the nonverbal communications (Savage & Rumbaugh 1977). We believe
that the cognitive base arises out of the maturing
chimpanzee's sensory-motor interactions with its
surroundings.

One product of the Lana Project (Rumbaugh
1977) has been an extensive study of conversation
(Gill 1977; Rumbaugh & Gill 1976a & 1976b). For
the purposes of the Project, conversation was defined as a linguistic type of exchange between two
beings, where there was both novel comment and response on the part of both participants. We are of
the opinion that the study of conversation serves
clearly to support the contention that the ape can
employ cognition in efforts to achieve at least
certain types of goals. The following conversation
(Gill 1977) illustrates the point that both the
content and the frequency of Lana's comments are
contingent upon the transmissions received by her
in the context of a conversation. All novel statements by Lana are marked by asterisks (**). The
conversation occurred during an experimental inquiry as to the variables which might influence the
probability of engaging Lana in a conversation.
The time was late afternoon, and the paradigm
called for Tim to ask Lana chimpanzee, through use
of his computer-based keyboard, what it was that
she "wanted to eat."

Tim: ? Lana want what eat.   16:23 p.m.

Lana: Lana want chow.   16:23

At this point, Tim put water into the milk
dispenser, an act which was called for by the paradigm so as to substitute a lesser-preferred liquid
(water) for the highly preferred food (monkey chow)
which Lana had requested. Lana observed the substitution but could not see the specific liquid
being put into the milk dispenser. That Lana noted
the event is certified by her next statement--

Lana: Please machine give milk.   16:28

This sentence (a stock one established through
operant procedures) resulted in Lana obtaining a

bit of the liquid, which she then drank. Had it
been milk, Lana would have asked for repeated rations of it. She did not, but continued--

Lana: ? You put chow in machine.   16:28

Tim: Chow in machine.   16:28

(This, too, was done in accordance
with the experimental paradigm, which
in this particular setting required
that Tim lie to her to see whether or
not she would agree.)

Lana: ? You put chow in machine.

Tim: Chow in machine.

Lana: ? You put chow in machine.

Tim: Chow in machine.

(Lana seemingly kept questioning Tim's
assertion that he had put chow in the
machine.)

Lana: ? Tim put juice in machine.   16:31

(Since no chow had been made available,
perhaps juice might be?)

Tim: Juice in machine.   16:31

(Once again, lying.)

** Lana: No juice.   16:32

(Here she was declaring, we believe,
that Tim's statement was not true;
there was no juice in the machine.
This was the first time Lana made this
statement.)

Tim: ? What in machine.   16:33

Tim: ? What in machine.   16:33

** Lana: Water in ... milk.   16:35

(She waited a long time and went to

various keys such as "machine," "milk,"
and "room" before concluding ... milk.
We believe that she was saying the best
she could that water was in where the
milk normally is. Also a novel statement.)

** Lana: ? You want water in machine. 16:35

(Here we believe that Lana wanted to
know if Tim really meant that water
should be there. Also a novel statement.)

Tim: No.   16:35

Lana: ? You move water behind room. 16:36

Tim: Water behind room.   16:37

(Which was a true fact and Tim was just
declaring it so.)

** Lana: ? You move water out-of   16:37
machine.

(A novel and totally appropriate request by Lana, given the context. This
statement was of recent origin, used
only a very few times prior to this
occasion.)

Tim: Yes.   16:41

(And he removed the water.)

Lana: ? You put juice.   16:45

Lana: ? You put juice in machine. 16:45

Tim: No Lana drink.   16:45

(It was time for her to eat chow.)

Lana: ? You put more juice.   16:46

Tim: No juice.   16:46

(You can't have any. He was not going
to give her any juice.)

Lana: <u>? You put juice</u> ...

(Tim cut her off.)

Tim got annoyed with her at this point because he
had declared that she could not have juice.

Tim: <u>No juice.</u>   16:46

Lana: <u>? You put chow in machine.</u>   16:47

Tim: <u>No "in"</u> ... <u>"on".</u>   16:47

(The actual loading of the chow is to
place it "on" a vending device.)

Lana: <u>? You put chow on machine.</u>   16:48

(It should be noted that Lana changed
the preposition in accordance with
Tim's insistence.)

Tim: <u>Yes.</u>   16:48

(And Tim loaded chow "on" the machine.)

Lana: <u>Please machine give piece of</u> 16:52
<u>chow.</u>

(repeatedly until supply exhausted)

Of particular significance in this conversation is the fact that upon seeing Tim put something
into the milk dispenser, Lana used the appropriate
stock sentence, "Please machine give milk period"
in order to get a sample of it. As stated, had it
been milk, the odds are overwhelming that she would
have asked for repeated rations of it until its
supply was exhausted. Next, she returned to her
request for chow. Failing to obtain any, for there
was none to be had — Tim's declarations to the contrary — she shifted tactics and asked for juice. In
response to Tim's insistence that there <u>was</u> juice,
when in fact there was none, Lana declared, "No
juice." We believe it is evident that through the
course of this representative conversation that
Lana tracked Tim's comments quite accurately and
responded with equal appropriateness. This would
not have been possible were it the case that Lana
responded only on the basis of specific past condi-

tionings. Lana's novel statements, "No juice,"
"Water in ... milk" (interpreted to mean that she
recognized that water was in the milk dispenser),
"? You want water in machine," and "? You move water out-of machine" are taken as evidence that Lana
exercised significant cognitive skills in the formulation of responses that were novel and correct
as well as apparently appropriate to the context
and the course of the conversation.

Köhler (1925) demonstrated a half-century ago
that chimpanzees, left to their own devices, will
see relationships between sticks and boxes and determine how to use them to gain access to food
which is otherwise beyond reach. Chimpanzees need
experience in order to do so (Birch 1945), but
they do not extrapolate in any simple linear way
from this generalized experience to the formulation
of creative and adaptive tool-use. They do so because their intellect and cognition processes information to the end of determining similarities
and differences. It is this complex, covert psychological process of discrimination, arrived at
through evolution, that can be used as the foundations of creative, linguistic behavior both in
man and his primate relatives.

References Cited

Birch, H.
    1946   "The Relation of Previous Experience to
            Insightful Problem-Solving" Journal of
            Comparative Pyshilogical Psychology
            38.367-83.

Bricker, W. A. and D. D. Bricker
    1974   "An Early Language Training Strategy" in
            R. L. Schiefelbusch and L. L. Lloyd
            (eds.), Language Perspectives: Acquisit-
            ion, Retardation, and Intervention, pp.
            431-68, Baltimore: University Park Press.

Carroll, J. B.
    1974   "Towards a Performance Grammar of Core
            Sentences in Spoken and Written English"
            in G. Nickel (ed.), Special Issue of
            International Review of Applied Linguis-
            tics on the Occasion of Bertil Malmberg's
            60th Birthday, pp. 29-49.

Fouts, R. S.
    1974   "Origins, Definition, and Chimpanzees"
            Journal of Human Evolution 3.475-82.

Gardner, B. T. and R. A. Gardner
    1971   "Two-way Communications with an Infant
            Chimpanzee" in A. M. Schrier and F.
            Stollnitz (eds.), Behavior of Nonhuman
            Primates, pp. 117-85, New York: Academic
            Press.

Gill, T. V.
    1977   "Conversations with Lana" in D. M.
            Rumbaugh (ed.), Language Learning by a
            Chimpanzee, New York: Academic Press.

Hayes, K. J., and C. Hayes
    1955   "The Cultural Capacity of the Chimpanzee"
            in J. A. Gavan (ed.), Nonhuman Primates
            and Human Evolution, pp. 110-25, Detroit:
            Wayne State University Press.

Hewes, G. W.
    1977 "Language Origin Theories" in D. M.
          Rumbaugh (ed.), Language Learning by a
          Chimpanzee, pp. 3-53, New York: Academic
          Press.

Kellogg, W. N.
    1968 "Communication and Language in the Home-
          raised Chimpanzee" Science 162.423-7.

Köhler, W.
    1925 The Mentality of Apes, London: Routledge
          and Kegan Paul

Krauss, R. M.
    1977 "Language and Communication" paper pre-
          sented at the Conference on Nonspeech
          Language, Gulf Shores, Alabama.

van Lawick-Goodall, J.
    1968 "The Behavior of Free-living Chimpanzees
          in the Gombe Stream Reserve" Animal Be-
          havior Monograph 1.161-311.

Maier, N. R. F.
    1929 "Reasoning in White Rats" Comparative
          Psychology Monograph 6.93.

Mason, W. A.
    1970 "Chimpanzee Social Behavior" in G. H.
          Bourne (ed.), The Chimpanzee, Vol. 2,
          Basel: S. Karger.
    1976 "Environmental Models and Mental Modes:
          Representational Processes in the Great
          Apes and Man" American Psychologists
          31.284-94.

Mason, W. A., R. G. Davenport, and E. W. Menzel
    1968 "Early Experience and the Social Develop-
          ment of Rhesus Monkeys and Chimpanzees"
          in G. G. Newton and S. Levine (eds.),
          Early Experience and Behavior, Spring-
          field: Thomas.

McGinnis, P. K.
    1972 "Patterns of Sexual Behavior in a Com-
          munity of Free-living Chimpanzees" un-
          published doctoral dissertaion, Darwin
          College.

Miller, J. F. and D. E. Yoder
    1974 "An Ontogenetic Language Teaching Stra-
          tegy for Retarded Children" in R. L.
          Schiefelbusch and L. L. Lloyd (eds.),
          Language Perspectives: Acquisition, Re-
          tardation, and Intervention, pp. 431-68,
          Baltimore: University Park Press.

Patterson, F.
    n.d. "The Gestures of a Gorilla: Language Ac-
          quisition in Another Pongoid Species"
          unpublished manuscript.

Premack, D.
    1971 "On the Assessment of Language Competence
          in the Chimpanzee" in A. M. Schrier and
          F. Stollnitz (eds.), Behavior of Nonhuman
          Primates, Vol. 4, New York: Academic
          Press.
    1977 "The Human Ape?" The Sciences 17(1).20-3.

Rumbaugh, D. M.
    1970 "Learning Skills of Anthropoids" in L.
          A. Rosenblum (ed.), Primate Behavior:
          Developments in Field and Laboratory
          Research, pp. 1-70, New York: Academic
          Press.
    1971 "Evidence of Qualitative Differences in
          Learning Among Primates" Journal of Com-
          parative and Physiological Psychology
          76.250-5.
    1977a Language Learning by a Chimpanzee: The
          Lana Project, New York: Academic Press.
    1977b "Language Behavior" in A. M. Schrier
          (ed.), Behavioral Primatology: Advances
          in Research and Theory, Vol. 1. Hills-
          dale: Lawrence Erlbaum Associates.

Rumbaugh, D. M. and T. V. Gill
    1976a "Language and the Acquisition of Lan-
          guage-type Skills by a Chimpanzee (Pan)"
          in S. R. Harnad, H. D. Steklis, and J.
          Lancaster (eds.), Origins of Evolution
          of Language and Speech: Annals of the
          New York Academy of Science. 280.90-123.
    1976b "The Mastery of Language-type Skills by
          the Chimpanzee (Pan)" in S. R. Harnad,
          H. D. Steklis, and J. Lancaster (eds.),
          Origins of Evolution of Language and
          Speech: Annals of the New York Academy
          of Science 280-526-78.

Savage, E. S. and R. Bakeman
1976 "Sexual Dimorphism and Behavior in Pan Paniscus" paper presented at the sixth International Congress of Primatology, Cambridge.

Savage, E. S. and D. M. Rumbaugh
1977 "Communication, Language, and Lana: A Perspective" in D. M. Rumbaugh (ed.), Language Learning by a Chimpanzee: The Lana Project, New York: Academic Press.

Terrace, H. and T. Bever
1976 "What might be learned from Studying Language in the Chimpanzee? The Importance of Symbolizing Oneself?" in S. R. Harnad, H. D. Steklis, and J. Lancaster (eds.), Origins of Evolution of Language and Speech: Annals of the New York Academy of Science 280.579-88.

Tolman, E. C.
1948 "Cognitive Maps in Rats and Men" Psychological Review 55.189-208.

Yerkes, R. and A. Yerkes
1929 The Great Apes, New Haven: Yale University Press.

7

Linguistic Capabilities of a Lowland Gorilla

Francine Patterson

The forma mentera of psycholinguistics has in recent years shifted from a reverence for Chomskian grammars innately imprinted and genetically unfolded to an appreciation of language as a cognitive process fundamentally similar to other intelligent behaviors. One way to reach an understanding of the basic cognitive processes involved in language is to examine its characteristics in a mode other than vocal. The approach being taken in this symposium is precisely this. Another method is to study its form and function in a species other than Homo sapiens, a method that underlies the second theme of this symposium. By switching modes or species we can sort out the characteristics essential to language from those that are merely consequences of its association with a particular mode or its use by a particular species.

This statement reveals one of my basic assumptions: human language is not a reduncant phrase — language is an ill-defined phenomenon in communication which may have many manifestations, including the possibility of nonhuman forms. A recent critic of the ape language experiments, Georges Mounin states that "... the criteria for human language ... cannot be obtained through mere generalization of its present features, but must be sought through a comparative ... study of all systems of communication, including animal communication" (1976:2).

What I am exploring in my work with the gorilla Koko are the parameters of — dare I say it — comparative pedolinguistics. It seems to me futile to argue as to whether or not several educated apes

"have" human language — at least until there is a
consensus on when a human has language. A more productive approach might be to chart the similarities
and differences in its form and function in child,
chimp, and gorilla, thus, discovering more about the
various paths the evolution of intelligence can take.

Project Koko began as the result of a fortunate
series of coincidences. In 1972, the Gardners lectured at Stanford on the acquisition of American
Sign Language by the chimpanzee Washoe, an event
which fired me with the ambition to become involved
in the examination of the language abilities of the
great apes. Shortly thereafter, a visit to the San
Francisco Zoo gave me my first glimpse of Koko, then
three months old. I immediately singled her out as
a likely subject for a language experiment, but the
Zoo intended to keep her with her mother. Dis-
appointed but undaunted, I enrolled in a course in
sign language, in the event that the opportunity to
work with an ape should present itself. Nine months
later, I discovered that Koko, a victim of malnutri-
tion and shibella, had been separated from her mother but had since recovered in the Zoo's nursery
and was a thriving 20 pound infant. When I again
asked for permission to initiate the sign language
study, the Director immediately granted my request.
The next day, July 12, 1972, I began work.

Modeling my study after Project Washoe, I held
the expectation that Koko would prove the peer of
Washoe in language acquisition, despite a literature
which depicted the gorilla as intractable, negativistic and intellectually disadvantaged in comparison with the chimpanzee. Robert Yerkes described
gorillas as aloof, independent, and at times ob-
stinate and negativistic: "in degree of docility
and good nature the gorilla is so far inferior to
the chimpanzee that it is not likely to usurp the
latter's place ... in scientific laboratories" (1925:
74). He noted that in direct contrast to chimpanzees, gorillas showed a low level of motivation and
a positive resistance to imitation. Similar conclusions have been drawn from more recent investigations of gorilla inteligence (Knoblock and
Pasamanick).

There is little question that the chimpanzee is
capable of conceptualization and abstraction that is
beyond the abilities of the gorilla. It is precisely
because of these limitations, which are apparently

genetically determined ..., that it is more difficult to work with them (1959:703).

My experience was quite different from what might have been expected following a reading of this literature. Koko was highly sociable (at times to a fault — her exuberant show-off behavior in the presence of visitors often proved disruptive to training which was initially carried out in the children's Zoo nursery in front of the public). She was also responsive to social reinforcers such as praise and tickling.

Although the gorilla learned signs most rapidly through the technique of molding (a process she initially resisted), Koko showed a substantial amount of imitation; in fact, her first recorded sign, DRINK, was a result of this process and not of molding. Before long, she was imitating domestic tasks, such as cleaning with a sponge, and at times I have had the experience of having my every activity and posture mimicked much to my amusement and sometimes to my embarrassment. Some days, Koko follows me like a shadow — writing, scrubbing the floor, using the phone, filing her nails, thoughtfully tapping her chin (cf. Figure 1).

Negativism has reared its ugly head on occasion but it is by no means a dominant characteristic of Koko's behavior. Incidence of negativistic behavior seemed higher when she was in the "terrible twos" and "fearsome fours" — but it has always been more a situation specific response than a generalized trait. One routine I devised to test Koko's comprehension of signs and relations involved asking her to perform a certain task as a condition of being allowed to come out of her room. For example, I would ask her to find the yellow ball, or the small shoe, or to put the baby under the blanket. Sometimes, the request was completely ignored, but more frequently she would bring me everything in the room except the requested item. Alternatively, she would put various objects, in, on, or under other objects, but, again, executing every permutation and combination except the correct one. It was an exasperating experience and I had the sneaking suspicion that she was more reinforced by our signs of frustration than by the end reward of her freedom. One way to gain her cooperation was to threaten to leave her completely along — suddently albeit reluctantly she made the correct response.

minuplation-sometimes same as children

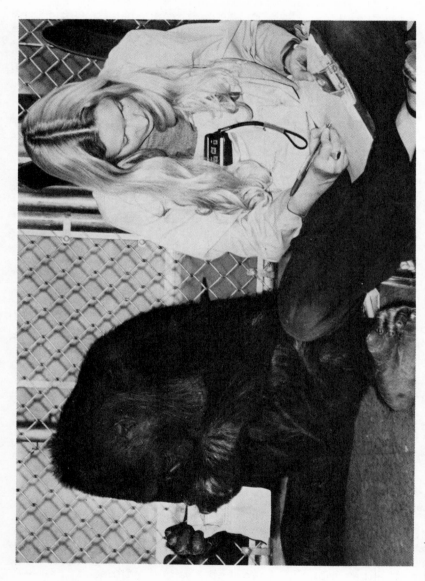

FIGURE 1. Koko's Imitative Behavior. Koko frequently imitates the behavior of her human companions. This and all photographs in this chapter are by Ronald H. Cohn, Ph.D.

At times, Koko's contrary behavior is more amusing than frustrating. For instance, she has signed FROWN or SAD, when asked to smile for a photograph. Occurrences such as this are informative in that they demonstrate her grasp of the concept of opposites.

Koko failed to match the gorilla stereotype in more ways than she fit it, however. Her acquisition of a large American Sign Language vocabulary and her performance in test situations, including standardized intelligence tests, call into question the widespread belief that the chimpanzee is the most intelligent of all nonhuman primates.

This brings us to the data on Koko that I would like to discuss in this paper, some of which is strictly comparative in nature. Initially, Project Koko was structured in form and intention quite similarly to Project Washoe in that I sought to investigate many of the same parameters: vocabulary development, generalization, semantic relations, comprehension, and productivity. My aim was to create a body of data from which direct comparisons could be made between Koko and Washoe and Koko and human children. Consequently, certain aspects of my methodology were similar: daily inventories of signs, double blind tests of vocabulary and comprehension, naturalistic observations on the use of sign, and studies of a behavioral development.

Vocabulary is a valuable index of cognitive and linguistic development. In fact, tests of vocabulary are considered to be the best single index of human intelligence. Koko's vocabulary is closely comparable in content and size to those of children and chimpanzees.

To be considered a reliable part of Koko's vocabulary, a sign must meet a criterion of spontaneous and appropriate use on at least one-half the days of a given month. This is somewhat different from the Gardner's (1969 and 1972) criterion (spontaneous and appropriate use for a period of 14 consecutive days) but it has seemed desirable to avoid daily drill sessions which might be imposed just for the sake of meeting such a criterion. Koko's sign vocabulary grew at the rate of approximately one sign per month during the first year and a half; by the end of January, 1973, 22 signs had met my criterion. This compares favorably with the progress

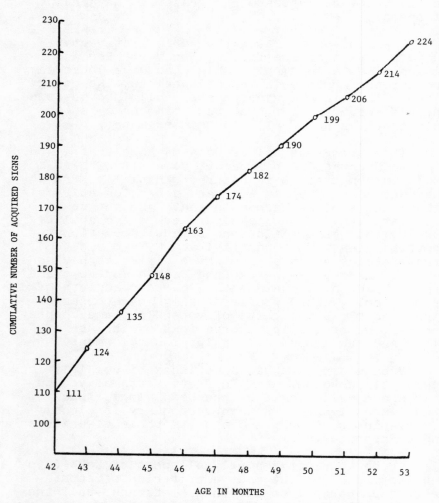

Figure 2. Cumulative number of signs in Koko's vocabulary
from age 42 to 53 months.

of Washoe, who acquired 21 signs during the first
18 months of her training. Since neither ape was
exposed to a sign language environment until she
was about one year of age, it is difficult to compare their vocabulary development with that of human
children who are exposed to their native langauge
from birth. However, despite her disadvantage, Koko
built a vocabulary of approximately 100 signs by the
time she was three and a half (30 months of training), a figure in the normal range for human children who build vocabularies ranging in size from 30
to 400 words between their second and third years.
For example, one report of a deaf child inventoried
132 signs at age 3 (Olson 1972). During the next
year, Koko's vocabulary doubled in size; at age four
and a half, she had approximately 225 signs at her
command (cf. Figure 2).

In order to directly compare Koko's vocabulary
development with that of Washoe, I applied the
Gardner's 14-consecutive-day criterion to Koko's
data. The Gardners report that after 36 months of
training, 85 of Washoe's signs had met their strict
criterion. At an identical point in her training,
112 of Koko's signs had met the Gardner's criterion.
Table 1 presents the vocabularies of Koko and Washoe in a manner that shows the considerable overlap between the two. Forty-six of the 112 signs
acquired by Koko within 36 months of training were
also acquired by Washoe in the same period of time.

Another figure available for comparison from
the published data on Washoe is the amount of signing at dinnertime. The Gardners started sampling
Washoe's signing in the 29th month of their project.
At this time, she used 50 signs during a 15 to 20
minute session taken at the evening meal. During
a one-hour dinnertime sample (5:00 PM to 6:PM) in
month 29 of this project, Koko used 251 signs.
(This would be approximately 63 every 15 minutes.)
While more direct quantitative comprisons will have
to await the release of additional data on Washoe,
these figures indicate that at similar points in
their training in sign language, the performance of
Koko and Washoe were closely comparable.

Words represented in the early vocabularies of
children (Nelson 1973) are **very similar** in semantic
content and proportion to those of Koko. For both,
the most prominent category is food and drink, followed by animals, then clothing and personal items.

TABLE 1. SIGNS IN KOKO'S VOCABULARY MEETING THE STRICT
GARDNER AND GARDNER CRITERION AS OF JULY 1, 1975 (approximately 3 years of training).

Signs acquired by Koko within 3 years of training not in
Washoe's vocabulary at the same point in her training.

alligator	earring	pinch-skin
apple	egg	potato
arm	feather	pour
around	fish	rubber
ask	fork	sandwich
bag-purse	grape	scratch
bean	hair	small
big	Koko	smile
bite	lipstick	soap
blanket	match	sock
bottle	milk	spice
bracelet	money	stamp
butter	mouth	straw
cabbage	necklace	tape
cake	nose	taste
candy	nut	teeth
cereal	on	there-that
chase	onion	thirsty
cold	orange	tiger
cookie	peach	time
craker	pick-groom	water
dry	pillow	

Signs in common with Washoe at same point in training.

baby	fruit	pen-write
banana	go	please
berry	good	quiet
bird	hat	red
book	hurry	ride
brush	key	sleep-bed
bug	light	sorry
cat	listen	string
catch	look	sweet
cheese	me	tickle
clean	meat	toothbruch
come-gimme	mine	tree
drink	more	up
eat-food	open	white
flower	out	wiper (bib)
		you

TABLE 1 continued.

Signs acquired by Washoe within 3 years of training not in Koko's vocabulary at the same point in her training.

black	funny	oil
car	grass	pants
chair	green	Roger
climb	Greg	shoes
clothes	hammer	smell
comb	help	smoke
cover	hug	spin
cow	hurt	spoon
dirty	in	Susan
dog	kiss	Washoe
down	leaf	Wende
Dr. G.	Mrs. G.	window
enough	Naomi	yours

The only category present in children's vocabularies but absent in both Koko's and Washoe's was that of places (e.g., pool, school). This category was the least frequently represented in the child lexicons and is understandably missing from those of the apes because of their more restricted environment.

Koko's vocabulary has expanded in every conceptual domain, with the greatest increase occurring in the object name category (cf. Table 2). This parallels the development of human children (Nelson 1973). Koko has acquired some words relating to the concept of time (e.g., now, time, finished) but she has not yet consistently used question words such as who, what, or why, although she can appropriately answer questions employing these words (1). Many children in the earliest stages of language acquisition do not use "wh-" words to express questions (Bowerman 1973). However, since the third month, Koko has used gestural intonation (retaining the hands in the sign position at the end of an utterance and seeking direct eye contact) to express questions (2). For example, one day in response to a woodpecker's tapping outside, one of Koko's companions, Barbara, signed, KOKO, LISTEN BIRD. Koko responded, BIRD? As she signed, she turned her head toward Barbara, cocked it, and opened her eyes wide — almost raised her eyebrows. Barbara signed YES, THAT IS A BIRD TAPPING OUTSIDE. LISTEN ... Then, Barbara tapped on the counter and pointed outside when the woodpecker made the noise again. This time, Koko signed, LISTEN BIRD, with a definite intonation.

As in the case of children and chimpanzees learning language, certain words are more difficult for Koko to articulate than others. Signs which require a great deal of manual dexterity are often simplified by the gorilla. For example, WATER and RUBBER are simplified from the "w" and "x" hand configuration to a forefinger extended from a loose hand or compact fist. Manual dexterity is not the only problems; some signs are physically impossible for Koko to articulate correctly (e.g., SAND and PURPLE) because of the small size of her thumb.

I have noticed a developmental change in the case of articulation of a certain group of signs in which the hands do not contact the body (non-contact signs) which parallels that reported for Washoe (Fouts 1973). See Peng (1976) for the details of

the classification of non-contact signs. Both Koko
and Washoe acquired non-contact signs more slowly
than contact signs before the age of 4. Prior to
the time when Koko could properly articulate non-
contact signs, such as FINISHED (cf. Figure 3) and
MILK, she tried to convert these signs into contact
signs by changing the place of articulation from a
point away from the body to one on the body.

Koko, like Washoe (Gardner and Gardner 1971)
and human children (Clark 1973), has spontaneously
generalized and overgeneralized her signs to novel
objects and situations. For example, the sign for
STRAW, learned initially with reference to drinking
straws, was spontaneously used to label plastic
tubing, a clear plastic hose (3/4 inch in diameter),
cigarettes, and a car radio antenna; OPEN, used
initially with reference to locked doors, was generalized to boxes, covered cans, drawers, and cup-
boards; NUT, learned as a name for packaged nuts,
was later used to refer to pictures of nuts in magazines, peanut butter sandwiches, roasted soybeans,
and sunflower seeds (but in this case only after
tasting some — she had first labeled them CANDY
FOOD); TREE, learned with reference to acacia
branches and celery, was overgeneralized to aspara-
gus, green onions, and other tall thin objects
presented vertically.

Not only has the gorilla demonstrated a capacity for acquiring a sizable vocabulary, she has
also spontaneously combined these words into meaningful and sometimes novel strings of two or more
signs. The first juxtaposition of gestures that
might be interpreted as a sign combination, GIMME
FOOD, was recorded on film during the second month
of the project when Koko was about 14 months old.
The signing was elicited by a drink held out of
reach, so an alternative interpretation might be
that this sequence was a combination of a natural
reaching gesture plus a sign. This second interpretation is supported by the finding that children
pass through a stage prior to the first joining of
words in which words and actions are combined to
express a statement (Greenfield and Smith 1976).
The sequence FOOD DRINK, elicited by a bottle of
formula, was also recorded during the second month,
but notes taken at the time state that Koko may have
been correcting herself, for we always referred to
the formula (cereal plus milk) as DRINK.

TABLE 2. 243 DISTINCT* SIGNS ACQUIRED[1] BY KOKO WITHIN 50 MONTHS OF TRAINING

TYPE	#	TOKENS
PROPER NOUNS:	4	Koko, Kate, Penny, Ron
PRONOUNS:	3	me, you, myself
ANIMATE NOUNS:	22	alligator, baby, bear, bird, bug, cat, clown, cow, dog, elephant, fish, frog, giraffe, gorilla, man, monkey, mouse, pig, rabbit, skunk*, tiger, moose
EDIBLE NOUNS:	46	apple, banana, berry, bean, bone, bread, butter, cabbage, cake, candy, carrot, cereal, cheese, cookie, corn, cracker, cucumber, drink*, egg, food*, fruit, grape, grass, gum, jello, leaf, lemon, lollipop-icecream, meat, medicine, milk, nut, onion, orange, peach, pepper, potato, pudding, raisin, salad, salt, sandwich, spice, tree, water
OTHER NOMINALS:	85	airplane, arm, ball, bag-purse, bed*, belt, bellybutton, blanket, brush, boat, book, bottle, bottom, box, bracelet, comb, cigarette*, chin, clay, clothes, drapes, ear, earring, eye, eyemakeup, finger, feather, flower, foot, fork, hat, hair, head, hurt*, injection, key, kiss*, knife*, leg, light, lighter, lip, lipstick, mask, match, mirror, mouth, nail, nailclipper, neck, necklace, nose, oil-lotion, pants, paper, pen-write*, shoulder, skin-pinch*, pillow, pimple, ring, rubber, shell, soap, sock, sponge, spoon, stamp, stethoscope, stomach, straw, string, swing*, sweater, tape, taste*, teeth-glass, telephone, time, toilet, tongue, toothbrush, train, whistle, wiper

TABLE 2 continued.

MODIFIERS: 34   all, bad, big, black, clean*, cold, different, dirty, dry*, good, green, hot, hungry, hurry, mad, mine, more, now, old, pink, quiet, rotten, red, sad, same, small, smoke*, sorry, stink*, sweet, thirsty, that-there, white, yellow

NEGATIVES: 5   can't, don't-not, don't know, no, nothing

VERBALS: 50   ask, bite, blow, break, catch, chase, clean*, come-gimme, cut*, do, draw, drink*, dry*, eat*, finished, frown, go, have, help, helpmyself, hide, hug-love, hurt*, kiss*, know, like, listen, look, make-fix, open, play, pick-groom, pinch-skin*, please, pound, pour, ride, sip, scratch, sit, sleep, smell, smile, swing*, taste*, think, tickle, turn-around, want, write-pen*

PREPOSITIONAL: 6   down, in, off, on, out, up

INTERJECTIONS: 2   darn, hi-bye

total: 5

1 Acquired indicates signs used spontaneously and appropriately half the number of days or more in a given month.

* Indicates signs occurring in more than one category because of overlap in usage. For example, "drink" can be both a noun and a verb, and the sign for "knife" (noun) is the same as the sign for "cut" (verb).

FIGURE 3. Koko's MILK. Koko signs MILK, a nontouch sign which is performed as if milking a cow with one hand. [Cohn]

During the third month of training, when Koko was 15 months old, she acquired the sign for MORE and soon began to use it spontaneously in conjunction with the signs for FOOD and DRINK. The first recorded combination using this sign was FOOD MORE for an additional serving of fruit. Washoe's first recorded combination of signs was GIMME SWEET at approximately 21 months of age. That this happens to be close to the age when human children begin to combine words may, however, be more of a coincidence than a parallel in development, because the Gardners' second chimpanzee subject, Moja, combined signs (GIMME MORE) at the tender age of 6 months (Gardner and Gardner 1975a). It might be noted that GIMME, which is one of the chimps' early combiners, is s gesture present in the repertoire of chimpanzees in the wild (van Lawick-Goodall 1968).

In the months following their first conjoining of words, human children exhibit only a few kinds of semantic relations in their utterances. Brown (1973) lists ten structural meanings that account for 75% of the first two-word sentences in child speech. Four of these are operations of reference (nomination, notice, recurrence, and non-existence) in which a constant term is joined with a variety of other words. The remaining six are relations (attribution, possession, location, agent-action, agent-object, and action-object) in which forms, such as verbs or nouns, are combined in several grammatical relationships. Table 3 lists all of these terms but one, NOTICE (to be discussed later) plus two other relations that occur somewhat less frequently in early child speech (person-affected dative, and dative of indirect object) along with utterances by Koko, Washoe, and deaf children which exemplify each relation.

It is evident that close parallels exist between Koko and the other subjects for each category. The correspondences extend to finer aspects of the data as well. For example, in both Stage I children and Koko, the possessive construction is limited to the expression of property-territory and part-whole relationships. While examples of the former relation, such as those given in Table 3, are frequent, the expression of part-whole relations is much less common in children and in Koko. BABY-HEAD, signed simultaneously for the head of a doll, is one of the few examples found in our samples.

TABLE 3. EARLY SEMANTIC RELATIONS EXPRESSED BY GORILLA
           KOKO, CHIMPANZEE WASHOE AND DEAF CHILDREN

Relation	Koko
Nomination:	that cat, that bird
Recurrence:	more cereal, more pour
Nonexistence:	me can't
Affected person,state:	sorry me, me good
Attributive:	hot potato, red berry
Genitive:	Koko purse, hat mine
Locative:	go bed, you out
Dative:	give-me drink
Agent-action:	you eat, me listen
Action-object:	open bottle, catch me
Agent-object:	alligator me (chase)

TABLE 3 continued.

Washoe[1]	Deaf Children[2]
that food, that drink	
more friut, more go	cookies more, please more
me can't	good finished
Washoe sorry	(me) angry, want open
drink red, comb black	that pink, red shoes
baby mine, you hat	Barry train
go flower, look out	Daddy work, (me) go home
give-me flower	give-me orange
you drink, Roger tickle	me eat, man work
tickle Washoe, open blanket	eat cracker, grab boy
	Daddy shoe (take off)

[1]EXAMPLES DRAWN FROM GARDNER AND GARDNER, 1971 and 1973.
[2]EXAMPLES DRAWN FROM KLIMA AND BELLUGI, 1972, AND SCHLESINGER AND MEADOW, 1971.

Koko has also made a number of the same mistakes in word usage as some children; for instance, she signed MORE CHEESE to request a first helping, and MORE SWING for another push on her trapeze. Her early use of MORE as a simple request form and her use of MORE in conjunction with verbs and even adjectives are patterns of overgeneralization reported in child language literature (Bloom 1970; Brown 1973).

Brown's (1973:40) finding that "about 75% of all utterances from all Stage I children studied ... are classifiable as expressions of this small set of semantic roles or relations" is one that may hold true for young apes learning language as well. Gardner and Gardner (1971) analyzed the 294 different two-sign sequences produced by Washoe during a 26 month period when she was reported to be between 2 and 4 years old, and found that a similar set of semantic relations accounted for 78% of the total. Using the categories listed in Table 3 plus an appeal category (e.g., UP PLEASE and HURRY GO), analogous to that in the Gardner and Gardner (1971) scheme, the two-sign combinations produced by Koko during 40 hours of samples in 1974 (8 hours each in the months of January, February, June, November, and December) were analyzed. These 12 semantic relations accounted for 75% of the 719 two-sign combinations (or 451 distinct utterance types). Thus, Koko resembles closely both Washoe and human children with respect to the early expression of relational meanings.

Although Koko has produced uninterpretable strings (as do some children) most of her utterances are well suited to the situation. Koko's linguistic interchanges with her companions indicate her comprehension of their utterances and illustrate her appropriate use of signs in a given context, as seen in the following example:

    K: YOU CHASE ME.
    B: MY NAME IS BARBARA.
    K: BARBARA CHASE.
    K: YOU TICKE ME.
    B: WHERE?
    K: TICKLE ARM.

Because sign language is used by Koko primarily in social situations in which she and her companions are cognizant of the context surrounding linguistic exchanges and have direct access to events and objects that serve as topics of conversation, it might be argued that her use of signs and our interpretation of them are dependent upon these extrinsic factors. That is, it is possible that Koko's use of sign langauge is not truly spontaneous, but elicited by cures and prompts that she receives from her companions. Double blind tests were administered to evaluate Koko's performance under conditions in which she had to respond linguistically to a stimulus that she alone could see. Lacking funds for a more sophisticated method, I employed the box test devised by the Gardners. A plywood box, 12 inches high by 14 inches wide by 14 inches deep, one side of which was a removable Plexiglass window, was used. One experimenter baited the box with a random selection from a pool of objects representing 30 of the nouns in her vocabulary, and then covered the box with a piece of opaque material. A second experimenter, who did not know the contents of the box, stood behind it and when Koko started the trial by uncovering the box, asked her what she saw (Figure 4). This test situation required a fair amount of discipline, and curiously enough we found that, like Washoe, Koko's interest and cooperation could be secured for no more than 5 trials a day and 2 sessions per week. Her methods of avoiding the task were varied — she would either respond to all objects with the same sign, refuse to respond at all, or regress to an earlier pattern of asking to have the box opened.

Despite the difficulties, Koko's responses on these tests were correct about 60% of the time. On a series of tests administered when Koko was 4 years old (September 1975), she gave the correct response on 31 of 50 trials (62%). This compares favorably with the level of performance reported for Washoe (54%).

Koko's errors on the box test were of four basic types (in order of frequency).

1. Requests to have the box opened (e.g., KEY BOX and KEY OPEN).

2. Conceptually related errors in which the response was for another item in the same class of objects (e.g., CANDY or CRACKER).

3. Perseverative errors (e.g., CLOWN on

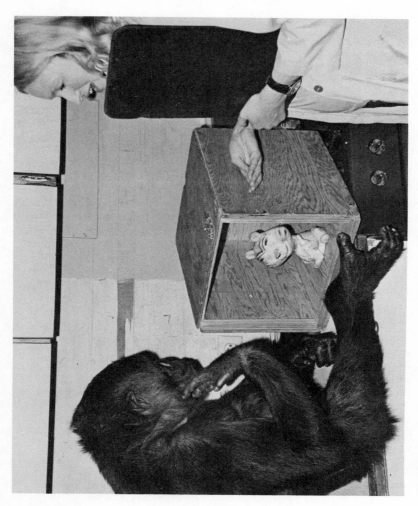

FIGURE 4. Correct Response. TIGER is a correct response given by Koko on a trial of the "box text." [Cohn]

two trials succeeding one on which it was
the correct response).

4.   Formally related errors in which the sign
     given was similar in articulation to the
     sign for the correct response (e.g., CLOWN
     for FLOWER).

On one interesting trial, Koko frowned, then
signed CAT RED when the box contained a red stereo-
scopic viewer. Upon examination, I found that the
slide in the viewer depicted a lion. Possibly, this
was an unrelated error; alternatively, she may have
been recalling what she had last seen in the viewer.
(Incidently, this response was scored as an error.)
Gardner and Gardner (1971) comment that many of
Washoe's errors in the double blind tests of her
nouns were errors of the second type, that is conceptually related.

In addition to the information such tests yield
on categorization and performance errors, they also
show that Koko's appropriate use of signs is not dependent upon cueing from her teachers and companions,
but is the result of an ability to retrieve and produce linguistic symbols spontaneously.

The usefulness of the form of double blind test
is limited, however. Only object names can be
tested and the procedure is expensive in terms of
time and personnel required, and often fails to
elicit the gorilla's cooperation.

Other techniques including the use of videotaped session in which experimenters with no knowledge of sign language are employed and the presentation of animate relational stimuli on film are
being devised and implemented.

It is also true that when the response required
in a test situation is verbal (as opposed to forced
choice) the probabilities of the subject emitting
the correct response by chance or as a result of
cueing of some kind by the experimenter are quite
small. For example, the probability of the gorilla
uttering the appropriate sign by chance, spontaneously or in response to a question, is about one
in 245, given a vocabulary of 245 signs. The
chances of her uttering a correct string of signs
by accident are even more remote. Therefore, I believe that there is even better evidence of Koko's

ability to retrieve, manipulate, and combine linguistic symbols appropriately in our records of her
day to day signing under completely naturalistic
conditions.

These records take the form of repeated samples
of Koko's signed utterances which have been routine-
ly recorded since week 15 of the project. Initially,
all linguistic exchanges between Koko and her companions, along with notes on the non-linguistic con-
test, were recorded during a 4 to 5 hour period
several times a week. As the gorilla began to produce increasing numbers of utterances, a decision
was made to limit sampling to 8 to 10 hours per
month. Each hour of the day between 9 o'clock and
5 o'clock was sampled once each month, so that
Koko's signing in the spectrum of her activities
during the day would be represented. Assignment of
sampling hours to days of the month was done on a
random basis. After the 40th week, these samples
were recorded on audio tape cassettes.

In order that a more accurate picture of the
pace and volume of Koko's signing during the course
of a day would be obtained, one 8-hour long sample
was taken once each month in addition to the eight
1-hour samples from month 12 of the project to the
present. Videotaped samples (30 to 90 minutes per
month) have been taken since the 16th month (November 1973). These will be invaluable in the analysis
of certain subtle characteristics of her signed utterances, such as form, intonation, and segmenta-
tion, that only a visual record can fully capture.

The following is an excerpt from an 8-hour
sample taken at the end of last month (January 31,
1977, 9:00 A.M.):

      (It is time for Koko's breakfast which normally
consists of a square of rice bread and a glass of milk).

P: WHAT DO YOU WANT?
K: APPLE DRINK.
P: HOW ABOUT ... (Penny is about to suggest her usual
     rice bread but Koko interrupts, signing
K: APPLE.
P: WHAT TIME IS IT? WHAT TIME? (Penny wants to
     elicit a new sign in her vocabulary, BREAKFAST.
     Koko is persistently requesting some apple juice
     — her favorite beverage which is not a usual
     breakfast item,

K: DRINK.
P: BREAKFAST.
K: BREAKFAST.
P: O.K., YOU WANT BREAKFAST?
K: GOOD HAPPY.
P: WHAT DO YOU WANT FOR BREAKFAST?
     (A rhetorical question — Penny is not really
     going to give her a choice, and gets out rice
     bread, saying [vocally] as she does so.)
P: I've got something good for gorillas ... right
     here's breakfast. (Penny offers rice bread to
     Koko.)
K: BOTTLE THERE APPLE. (Koko indicates a bottle of
     apple juice in the refrigerator.)
P: KOKO? (Penny, who does not intend to alter the
     breakfast menu, firmly closes the refrigerator
     door, gets the rice bread out of the bag.)
K: MEAN ... (Koko had never before used this sign
     and Penny did not comprehend her utterance at
     the time, but recorded its configuration. Koko
     seemed to be trying to clarify her own intentions.)
P: WHAT'S THIS?
K: BOTTLE APPLE ... NICE THERE (points to her palm).
P: KOKO, YOU ARE SIGNING WRONG THINGS. (Penny wants
     Koko to sign for the rice bread but Koko rephrases
     her request.)
K: PLEASE MILK, PLEASE ME LIKE DRINK APPLE ... BOTTLE.
P: HOW ABOUT FIRST BREAKFAST?
K: BREAKFAST EAT SOME COOKIE EAT. (Now Koko has
     either made an error or another unacceptable re-
     quest — gorillas, like children, do not get sweets
     for breakfast.)
P: NO, NO, BREAD. (Penny molds BREAD and gives it to
     Koko. A glass of milk will complete Koko's break-
     fast. Penny has a glass of water out for powdered
     milk; Koko touches it and, then, the container of
     milk powder.)
K: HURRY DRINK MILK.
P: YES, PENNY MAKE?
K: GLASS ALL MILK. (Perhaps Koko is suggesting that
     Penny put all the milk powder in her glass, Penny
     molds MAKE and makes it.)
P: O.K., NOW WHAT?
K: TIME MILK. (Koko is allowed to drink the glass of
     milk. Koko, across the counter, hands Penny the
     empty glass and looks at a new doll a few feet
     away.)
K: HURRY GIMME.
P: WHAT? (Koko reaches for the new doll, but Penny
     stops her from getting it.)
K: BABY COME-ON HAPPY. (Penny picks up the doll.)

```
P:    THIS IS BABY ALIVE.
K:    LIKE BABY NEW.
P:    YES!  (Penny gives the doll to Koko.)
K:    GOOD HAPPY.  (Koko lies down on floor the doll in
      her arms signing,)
K:    SLEEP.  (Koko seems to have forgotten all about
      the apple juice ...)
```

In addition to these comprehensive samples, I have also been gathering data on Koko's responses to questions. This aspect of the study is particularly important, because it can provide information on Koko's understanding of abstract concepts, such as casualty and time (abilities so far not well documented in apes) as well as evidence for a grasp of major sentence constituents, an ability demonstrated by Washoe (Gardner and Gardner 1975). Again, desiring to obtain data from which direct comparisons between Koko and Washoe could be made, I employed a methodology similar to that of the Gardners. In order to examine the full spectrum of Koko's abilities, however, I included a wider variety of question frames including Why and How, and two additional modes, spoken English and simultaneous communication, or the use of sign accompanied by speech.

This study is still in progress and will be the subject of a future paper, but it introduces the second major point I wish to make: although similarities in approach and methodology were important for comparative purposes, from its inception, Project Koko diverged from Project Washoe in a number of significant respects.

1.   Deaf personnel have been employed as teachers and companions for the gorilla from the beginning of the project.

2.   The use of spoken English in the presence of Koko has been allowed and the methodology has provided for samples and tests of the gorilla's production and comprehension under conditions in which the two modes (sign and English) are separated.

3.   Complete records of the gorilla's signed utterances along with a description of the context in which they occurred have been kept so that analyses of sign order can be performed.

4. No attempt has been made to limit the size or control the content of Koko's vocabulary.

5. Secondary examinations of the gorilla's intelligence featuring the use of standard infant and pre-school intelligence tests and Piagetian techniques have been carried out on a regular basis, thus providing data on her cognitive abilities directly comparable in a quantitative way to those available on human children and apes.

I made no attempt to eliminate spoken language from the gorilla's environment, in part because the conditions of the facility housing of the project for the first year, San Francisco Zoo's children's nursery, precluded the possibility of such control. Instead, I decided to turn this situation to advantage by adopting a method known as simultaneous communication, or the use of American Sign Language accompanied by spoken English. I provided Koko with native signers as teachers and companions on a day to day basis and used sign language as the primary mode of communication because it would yield data most directly comparable to that available on the language acquisition of both child and chimpanzee. Speech was used as a secondary mode of communication. Early research had shown that although apes could learn to produce fewer than a half dozen spoken words, even after years of training, the animal's comprehension skills were in all cases far superior to their production skills. Koko had lived for a time in the home of the Director of the Children's Zoo and when I began to work with her, she was already responsive to several spoken words. We hoped that Koko would comprehend a considerable amount of English, even though she probably would not come to produce speech to a significant extent. Furthermore, it seemed possible that the redundancy of information conveyed by the two modes might facilitate learning (Ferster 1964). In short, the use of simultaneous communication would allow for the study of any receptive and productive skills the gorilla might develop in the vocal as well as the manual mode and for the possibility of the transfer of information between modes. In addition, this method of teaching might allow an evaluation of the validity of claims, such as the following: "apparently it is the child's innate capacity for auditory analysis that distinguishes him from the

FIGURE 5. Assessment of Children's Language Comprehension Test. Examples of three (top) and four (bottom) critical element items from the Assessment of Children's Language Comprehension Test. From Rochana Foster, Jane Giddan, and Joel Stark, 1973. Courtesy of Consulting Psychologists Press, Inc., Palo Alto, California.

chimpanzee" (Hebb, Lambert, and Tucker 1974:153).

The gorilla's responsiveness to spoken requests· in our daily interaction with her suggested that she did possess a considerable capacity for complex auditory analysis. She would often surprise us by translating English words and phrases into signs during the course of the day. For instance, once a visitor asked Koko's companion what the sign for GOOD was. Before an answer was given, Koko demonstrated the sign for GOOD to the visitor. On another occasion, Koko's companion was making her a sandwich and asked if she wanted a taste of butter (in English only, since her hands were occupied) and Koko responded TASTE BUTTER. She will also frequently respond to spoken suggestions to put refuse into the garbage, kiss a companion, find a towel, etc.

/The problem with evidence of this type is that most speech to the gorilla (as to children) occurs in situation in which a good guess on her part will yield the correct interpretation of a sentence and/ or an appropriate linguistic response./

To circumvent this difficulty, I administered a standardized test to assess the extent of Koko's comprehension of English, sign, and simultaneous communication early in 1976, when Koko was 4 1/2 years old. The test, "The Assessment of Children's Language Comprehension" (Foster, Giddan, and Stark 1973), consisted of 40 large cards, 6 inches by 14 inches on which were printed 4 to 5 black-and-white line or silhouette drawings representing objects, attributes, or relationships between objects (cf. Figure 5). The first 10 cards, used to test single vocabulary items, depicted 5 objects each for a total of 50 vocabulary items. The remaining 30 cards tested comprehension of phrases, consisting of from 2 to 4 critical elements (e.g., little clown; little clown jumping; happy little clown jumping). Ten items at each of these 3 levels of difficulty were administered under three conditions (sign only, English only, and sign and English simultaneously) for a total of 90 trials. All trials were recorded on videotape and the sign-only and voice-only trials were alternated (so that these two conditions were tested simultaneously) and administered blind. That is, a second experimenter independently pre-selected the target picture for each trial, and compiled a list of items corres-

TABLE 4. KOKO'S PERFORMANCE ON THE ASSESSMENT OF CHILDREN'S LANGUAGE COMPREHENSION TEST

# Critical Elements	Chance	Sign +Voice %	Sign %	Voice %	Total %
Vocabulary--one (50 items)	20	72			
Two (e.g., happy lady)	25	70	50	50	56.7
Three (e.g., happy lady sleeping)	25	50	30	50	43.3
Four (e.g., happy little girl jumping)	20	50	50	30	43.3
	═══	═══	═══	═══	═══
Totals (two, three and four elements)		56.7	43.3	**43.3**	47.7

ponding to the numbered cards (which were arranged
in order of level of difficulty). The person ad-
ministering the test followed the list without
looking at the cards (which were kept face down until trial time) and presented them in such a way
that only the video camera and the gorilla could
see the stimulus pictures. After Koko had responded
by pointing to one of the pictures, the experimenter
looked at the card to determine whether or not her
response was correct and rewarded her accordingly.

Koko performed significantly better than chance
under all three conditions and at all levels of difficulty (cf. Tables 4 & 5). I had anticipated that
comprehension might be enhanced under the simultaneous communication condition. Although there was
a trend in this direction, it was not statistically
significant. The data do clearly indicate that Koko
comprehends novel statements in sign language and
spoken Enlish with equal facility. An unexpected
finding was that Koko's performance at all levels
of difficulty was quite similar. She responded
correctly to 43% of the 30 most difficult items (4
critical elements), a level of performance that
matches the norm for educationally handicapped
children four to five years old. Her performance
on the two critical element items was slightly better than her performance on items at the next two
levels of difficulty, but not significantly so.
This seems to indicate that some factor other than
complexity of the phrase was operating to limit her
performance. Evidence from her behavior during
testing sessions indicated that motivation may have
been the limiting factor. From session to session
and even from trial to trial, Koko showed wide variations in her level of motivation to do the task.
Her best performance was obtained on the first
phase of the test (simultaneous communication, 2
critical elements). The gorilla's level of motivation was also highest at this stage — the test
was novel and she quickly completed all 10 problems
in one session. In the subsequent phases of the
experiment, her performance frequently fell below
chance in sessions consisting of more than 5 trials.
After responding correctly on the first few trials
of a session, Koko would lapse into a series of consecutive errors. She seemed to use incorrect responses as a way of signalling that she had had
enough testing — responding rapidly without considering each of the alternatives (3).

TABLE 5. RESULTS OF CHI-SQUARE TESTS ON KOKO'S LEVEL OF
PERFORMANCE ON THE ASSESSMENT OF CHILDREN'S LANGUAGE COMPRE-
HENSION TEST.

Variable	$x^2(1)$	Significance Level
Sign + Voice (2,3 & 4 critical elements)	18.6	.001
Sign Only (2, 3 & 4 elements)	6.7	.01
Voice Only (2, 3 & 4 elements)	6.7	.01
Two Critical Elements (all conditions)	16.0	.001
Three Critical Elements (all conditions)	5.4	.025
Four Critical Elements (all conditions)	10.2	.005
Sign + Voice versus Sign Only (all levels)	2.1	ns
Two versus three Critical Elements (all conditions)	2.1	ns

I have encountered this difficulty when repeating other standardized tests — once Koko is familiar with the materials and the novelty is gone, she may either refuse to do the task or her performance may deteriorate. The standard laboratory phenomena of learning to learn and practice effects would lead one to expect improvement under these conditions. Paradoxically, it may be that an exposition of the gorilla's intellectual capacities will prove to be a test of our ingenuity in devising stimulating new tasks.

A number of innovations have marked Koko's progress over the past 4 1/2 years: She has invented signs and names for novel objects; she talks to herself; she engages in imaginative play using sign; and she has used language to lie, to express her emotions, and to refer to things displaced in time and space.

The first evidence of her innovative ability occurred very early in the study. Koko produced gestures which closely resembled the signs for COME or GIVE-ME, GO, HURRY, and UP. These were distinctive, because they appeared without direct training at a time when the great majority of Koko's signs were being acquired by molding. Because these same gestures have been noted in observations of other young gorillas in captive conditions, who have not been exposed to sign language, they may be part of the gorilla's natural repetoire of gestures (4). Other gestures which resemble signs, but which have not been observed in untrained gorillas, have been spontaneously produced and used productively by Koko. These signs seem to qualify as true "inventions," for they are not standard forms in Ameslan. Examples are: BITE, TICKLE, STETHOSCOPE, and DARN.

The sign for BITE is done by clamping one clawed hand onto the side of the other hand held flat, palm down. Koko's sign for BITE (open mouth contacting the side of the hand held palm down) originated quite dramatically one afternoon: Koko had play-bitten Ron, one of her companions, just a little too hard and once too often and he bit her back on the knuckles. Taken by surprise, and perhaps even a little shocked, she came to me for comfort. When I asked, WHAT HAPPENED? Koko cast a woeful glance at the offender and placed her mouth on her hand — the meaning was unmistakable. She

Figure 6. The Sign of DARN. DARN is Koko's invented
expletive. It is performed by hitting the back of
a hand in a clenched fist onto a surface or object.
Note the facial gesture accompanying the sign —
lips drawn in, mouth compressed — which is a natural
sign of annoyance in both gorillas and humans. [Cohn]

has continued to use this form of the sign appropriately (even in imitation of our correctly formed version of the sign) ever since.

Since there is no Ameslan sign for TICKLE, I adopted the sign used by the Gardners, which is executed by drawing the index finger across the back of the hand. Though her companions have repeatedly modeled and molded this sign, Koko has consistently used a more iconic gesture of drawing her index finger across her underarm which I have come to accept as a variant of TICKLE sign with a perhaps more logical place of articulation.

I learned the sign for STETHOSCOPE (index finger of one hand to the ear; the other hand in a fist on the chest) only after Koko invented her own (in which she placed an index finger to each ear).

One interesting invention of Koko's is a sign done by hitting the back of a tight fist onto a surface or object (cf. Figure 6). This gesture closely resembles a natural killing motion observed in chimpanzees but which has not been reported in gorillas in the wild (as far as I know). Koko uses the sign to express herself in situations in which her goals have been frustrated in some way or in which she seems annoyed by something, and I have translated it as DARN. For instance, she signed DARN BIRD BIRD, while a bird outside was giving a cry resembling a distress call for several minutes (it seemed like hours — I had tried and failed to locate the bird).

Two other gestures invented by Koko are less clearly iconic. One can only be defined as meaning "walk-up-my-back-with-your-fingers" (both hands are placed palm up on the floor, behind the back). This gesture is not often used in conjunction with other signs and has not been included in Koko's vocabulary. The other gesture had me completely mystified for some time. It is executed by quickly stroking the index finger across the lips and is preceded or followed by a noun or the pointing gesture, THAT. It is not directed to companions but occurs when Koko is looking through magazines, playing by herself with toys, nesting, or noticing an object in passing. When forced to give it an English equivalent, I called it NOTE, and later realized its resemblance to the construction "hi" & noun found by Bloom (1970) which is used by

children not as a greeting but in taking notice of
the presence of an object.

Another creative aspect of Koko's use of sign
language is her practice of varying the place of
articulation of several action signs, such as
TICKLE, PINCH, or PICK, to convey slight differences
in meaning. (See the characteristic of directionality mentioned by Peng in this volume.) At times,
she will sign TICKLE on her inner thigh instead of
her underarm or on both underarms (as a kind of emphasis). The location of the PINCH sign is highly
variable: It may be done on her leg, stomach, arm,
neck, etc., rather than on the back of the hand, indicating exactly where she wants to be pinched. The
sign, PICK, is properly done on the index finger,
but Koko signs it on her teeth when she needs help
dislodging food stuck between her teeth, or as a
request for dental floss. In addition, Koko has
signed SCRATCH on her back (to request a back
scratching) and on her finger (when confronted with
a recently inflicted scratch wound on her companion's finger), as well as on the back of the hand,
which is the correct form.

These variations are used consistently, are
readily interpretable, and seem to be deliberate
and efficient alterations made by Koko to express
her desires. I have observed another use of sign
with this simultaneous aspect (cf. the characteristic of simultaneity in Peng's article) not only in
Koko's but also in my own signing which seems for-
tuitous.

Two distinct signs are merged into one by combining the place of articulation of one with the
hand configuration of the other. An example produced by Koko was CHUCK-TOILET — a C done on the
nose. The hand configuration was that of the proper name but the location was that of the toilet
sign. This occurred after Chuck had emptied her
potty. I have done similar gestures unintention-
ally, usually when I am signing rapidly (e.g.,
APPLE-DRINK with the drink sign done on the side of
the face instead of at the mouth). This phenomenon
resembles Hockett's (1959) description of blending,
a process exemplified by the word "slithy," a combination of "lithe" and "slimy" (5). Hockett
stated that such words can be created by this process without conscious planning, and hypothesized
that the roots of productivity lie in such forms;

"... in (our) prehuman ... ancestors ... (when) a
few blends were communicatively successful ... the
closed circle was broken and productivity was on its
way" (1959:37). Signs such as PREFER may have originated in this way — the place of articulation is
that of the sign for LIKE and the hand configuration
is that of the sign for BETTER.

In addition to creating new signs, Koko invents
names for new objects in her environment, some of
which are strikingly apt. For example, she called
a ring a FINGER BRACELET, a zebra a WHITE TIGER, a
mask an EYE HAT, and a Pinocchio doll an ELEPHANT
BABY. Koko also spontaneously comments on her human
companions about occurrences in her surroundings.
For example, she signed LISTEN QUIET when an alarm
clock stopped ringing in the next room, and SEE BIRD
when she saw a picture of a crane in a stereo viewer.

Not all of Koko's utterances are directed to
her human companions, however. She has been observed talking to herself, her dolls, various animals, and to another gorilla. Her habit of talking
to herself, that is, using signs meaningfully but
directing them to no one but herself, has increased
in frequency over time and the utterances have gradually increased in length as well. Often, while
nesting in clothes or blankets and toys, she will
stop arranging the nest momentarily and sign THAT
RED (indicating a piece of red cloth), or ME SLEEP
and lie down in the nest. Frequently, while look-
king through magazines, she will comment on herself
about what she sees; for example, THAT TOOTHBRUSH
and THAT NUT. Recently, I noticed Koko signing
THAT KOKO to a picture of King Kong depicted on a
bowl I bought for her at the supermarket. At times,
Koko seems embarrassed when her companions notice
that she is signing to herself, especially when it
involves her dolls and animal toys. The other day,
while Koko was signing KISS, after kissing her alligator puppet, I caught her eye. She abruptly stopped signing and turned away. At times, when her
companions have been otherwise occupied, Koko has
taken her toy gorillas into the room furthest from
them and engaged in sign by herself while playing.
One day, she seemed to structure an imaginary social
situation between two gorilla dolls. She placed
the gorillas before her and signed BAD, BAD, while
looking at the pink gorilla, then, KISS to the blue
gorilla. Next, she signed CHASE TICKLE and hit the

two of them together (making them play?), then
joined in and wrestled them both at once. When the
play bout ended, she signed GOOD GORILLA, GOOD,
GOOD." At this point, Koko noticed that Cathy, her
teacher, was watching and left the dolls.

This observation and others like it have
strengthened my expectation that, given the opportunity, Koko would sign to other gorillas and perhaps even teach them. In his recent review, Mounin
(1976) asserted that Washoe did not have language,
because she did not use it to lie or to convey what
she had learned to her conspecifics. There is preliminary evidence that Koko does both of these
things.

For the past several months, Koko has had contact with Michael, a 3 1/2-year old male gorilla
with whom she now shares the project trailer. Almost as soon as they met, the gorillas were signing
COME to each other through the fencing that se-
parates their rooms. During their play sessions
together, Koko frequently signs to Michael. CHASE
is a common request and Michael, who has just acquired this sign, will often respond by pursuing
Koko. She has also asked him to do such things as
COME TICKLE-FOOT. One day, she signed ME HIT YOU,
then, followed up her threat by initiating a play-
fighting bout with a few blows. Michael has used
his small but growing vocabulary of signs to communicate with Koko as well.

Although it is difficult to empirically demonstrate intent, Koko has made statements in response
to questions about her misbehavior which appear to
be lies. For instance, once she was caught in the
act of trying to break a window screen with a chop-
stick she had stolen from the silverware drawer.
When asked what she was doing, Koko replied SMOKE
MOUTH and proceeded to place the stick in her mouth
as though she was smoking it (this is a game we engage in frequently with sticks and other cigarette
shaped objects). On another occasion, Koko, who
had just tipped the scales at 90 pounds, sat on the
kitchen sink and it sank about 2 inches. Not knowing how it had happened, I asked Koko, DID YOU DO
THAT? and Koko signed KATE THERE BAD, pointing to
the sink. Kate, my deaf assistant who had witnessed the incident, defended herself by explaining
the situation.

Koko's apparent prevarications usually take place under interrogation at the scene of the crime, immediately following the misbehavior. However, Koko will often respond to questions about such incidents long after their occurrence, indicating that she is capable of using language to refer to events and objects removed in time and space. This ability, known as displacement, is considered to be a fundamental characteristic of human language (Hockett 1960; Bronowski and Bellugi 1971). For example, the day after Koko bit a companion, I asked her WHAT DID YOU DO YESTERDAY? She replied, WRONG, WRONG. WHAT WRONG? I queried. BITE. Data being collected in the question-answer study give further evidence of this ability:

Q: WHAT DID I SAY YOU COULD HAVE AFTER BREAKFAST?
A: COOKIE (An accurate response; the promise was made about an hour earlier.)
Q: DO YOU REMEMBER WHAT HAPPENED THIS MORNING?
A: PENNY CLEAN (Not the answer I expected, but again accurate. Koko had been upset by something prior to my arrival that morning and had made a mess of her room. It took me the better part of an hour to clean it.)

The following conversation took place three days after the event discussed:

P: WHAT DID YOU DO TO P?
K: BITE.
P: YOU ADMIT IT? (Previously, Koko had referred to the bite as a scratch.)
K: SORRY BITE SCRATCH (P shows Koko the mark on her hand — it really does resemble a scratch.)
K: WRONG BITE.
P: WHY BITE?
K: BECAUSE MAD (A few moments later, it occurred to P to ask Koko,)
P: WHY MAD?
K: DON'T-KNOW.

The preceding example is noteworthy, because Koko makes reference to a past emotional state, her anger, without actually experiencing it at the moment. This is a clear indication that she is able to separate affect from the context of her utterances, another important feature of displacement.

There is other evidence that Koko can reflect upon and report about her feelings. She has spon-

taneously informed her companions that she is happy
or sad or tired and regularly answers questions,
such as HOW ARE YOU? and HOW DO YOU FEEL? The gorilla has even reported about her fears: Koko has
always been repulsed by lizard-like creatures and
toys. One day, several hours after a play session
in which she avoided direct contact with the gorilla, Michael, I asked her if she was afraid of him.
She made no response, so I rephrased the question.

P: WHAT ARE YOU AFRAID OF?
K: AFRAID ALLIGATOR.

What can be concluded from this brief and partial overview of Project Koko? As an early and
partial answer to this question, I would like to
reformulate a statement made by Noam Chomsky in
1968, by substituting the word primate for "human"
and the phrase linguistic communication for "unique
human possession":

> ... There is no better or more promising way to
> explore the essential and distinctive properties of
> primate intelligence than through the detailed (com-
> parative) investigation of the structure (and function-
> ing) of ... linguistic communication.

Notes

*This research was made possible by grants
from the Spencer Foundation and the National Geographic Society and support from the recently established Gorilla Foundation of Menlo Park, California. I would like to acknowledge the San Francisco Zoo for making the gorilla Koko available and
Professors Richard C. Atkinson and Karl H. Pribram
for their invaluable support. I am especially
grateful to Ronald H. Cohn and Barbara F. Hiller
for their continued assistance with the project.

1. In the following example, taken from the
diary records, Koko's companion described the in-
terrogative intonation in a way very similar to
Fischer's description of it for the deaf: "Yes-no
questions are marked by a raising of the eyebrows
and an expectant look on the face."

2. Responses to Why questions have not yet
been reported for chimpanzees learning language.
Koko now regularly answers such inquiries. For
example, one day, a deaf companion pursued her

most cherished ball under the trailer in a game of Keepaway, and gained possession of it in violation of Koko's special rules (which I had failed to adequately explain to her opponent). Koko's response was to give her companion a play bite on the posterior. About 20 minutes later, when asked why she bit her friend, she replied, HIM BALL BAD.

<u>3</u>. One other interesting source of errors, not due simply to inability to respond correctly, and perhaps due to her boredom with the task, was the following. In some instances, Koko would attend to the instructed phrase, and even after correctly translating or imitating the signs in the instruction, sign about another oject pictured on the card, and then point to the item containing that object. For example, when given the instruction, find BALL UNDER TABLE, Koko signed CAT and pointed to the cat under the table. Making up her own rules?

<u>4</u>. However, Dian Fossey (personal communication) has not observed the "hurry" gesture in free-living mountain gorillas.

<u>5</u>. Note that there is a significant difference between Patterson's description of the combination of two signs, such as CHUCK-TOILET and APPLE-DRINK, and Hockett's description of the combination of "lithe" and "slimy" for "slithy. The former is a silmultaneous combination, in the sense described by Peng in this volume, whereas the latter is a linear combination — Editor.

References Cited

Bloom, L.
   1970   Language Development: Form and Function
          in Emerging Grammars, Cambridge:
          The M.I.T. Press.

Brown, R.
   1973   A First Language, Cambridge: Harvard
          University Press.

Bronowski, J. S. and U. Bellugi
   1970   "Language, Name, and Concept" Science
          168.699-73.

Clark, E. V.
   1973   "What's in a Word? On the Child's

Acquisition of Semantics in His First Language" in T. E. Moore (ed.), <u>Cognitive Development and the Acquisition of Language</u>, New York: Academic Press.

Ferster, C. G.
1964 "Arithmetic Behavior in Chimpanzees" <u>Scientific American</u> 210.98-104.

Foster, R., J. T. Gidden, and J. Stark
1973 <u>Assessment of Children's Language Comprehension</u>, California: Consulting Psychologists Press, Inc.

Fouts, R. S.
1973 "Acquisition and Testing of Gestural Signs in Four Young Chimpanzees" <u>Science</u> 180.978-80.

Gardner, R. A. and B. T. Gardner
1969 "Teaching Sign Language to a Chimpanzee" <u>Science</u> 165.644-72.
1971 "Two-way Communication with an Infant Chimpanzee" in A. M. Shcrier and F. Stollnitz (eds.), <u>Behavior of Nonhuman Primates</u>, Vol. 4, New York: Academic Press.
1972 "Communication with a Young Chimpanzee: Washoe's Vocabulary" in R. Chauvin (ed.), <u>Modeles Animaux du Comportement Humaine</u>, 198.241-64, Paris: C.N.R.S.
1975a "Early Signs of Language in Child and Chimpanzee" <u>Science</u> 187.752-3.
1975b "Evidence for Sentence Constituents in the Early Utterances of Child and Chimpanzee" <u>Journal of Experimental Psychology</u> 104(3).244-67.

Greenfield, P. M. and J. H. Smith
1976 <u>The Structure of Communication in Early Language Development</u>, New York: Academic Press.

Hebb, D. O., W. E. Lambert, and G. R. Tucker
1974 "A DMZ in the Language War" in James B. Maas (ed.), <u>Readings in Psychology Today</u>, pp. 152-7, California: Ziff Davis.

Hockett, C. F.
1959 "Animal 'Language' and Human Language" in J. N. Spuhler (ed.), <u>The Evolution of</u>

Man's Capacity for Culture, pp. 32-8, Detroit: Wayne State University Press.
1960 "Origin of Speech" The Scientific American 203.88-96.

Knoblock, H. and B. Pasamanick
    1959 "The Development of Adaptive Behavior in an Infant Gorilla" Journal of Comparative and Physiological Psychology 52. 699-704.

Mounin, Georges
    1976 "Language, Communication, Chimpanzees" Current Anthropology 17(1).1-21.

Nelson, K.
    1973 "Some Evidence for the Cognitive Primacy of Categorization and its Functional Basis" Merrill-Palmer Quarterly 19.21-40.

Olson, J. R.
    1972 "A Case for the Use of Sign Language to Stimulate Language Development During the Critical Period for Learning in a Congenitally Deaf Child" American Annals of the Deaf 117.397-400.

van Lawick-Goodall, J.
    1968 "A Preliminary Report on Expressive Movements and Communication in the Gombe Stream Chimpanzee" in P. C. Jay (ed.), Primates: Studies in Adaptation and Variability, pp. 313-74, New York: Holt, Rinehart, and Winston, Inc.

Yerkes, R. M.
    1925 Almost Human, New York: Century.

8

Linguistic Potentials of Nonhuman Primates

Fred C. C. Peng

The study of language acquisition in apes, with
its comparative pedolinguistic approach beginning
with the Hayes and the Gardners, has traditionally
been concerned with the cognitive and the linguistic aspects. The ape-language papers assembled in
this volume are illustrative of this hard-earned
and good tradition, one that must be commended for
its place in the advancement of science. Moreover,
there is no denying that the continued pursuit of
language acquisition in apes along these lines has
a lasting scientific value. But it is also true
that relatively little attention, except for works
like Lieberman (1972), has been paid to the anato-
mical/physiological aspect of the subject, although
neurolinguists have lately become interested in this
area of research. The present article is offered
to somewhat balance the current views in comparative
pedolinguistics.

In line with this reasoning, my purpose in this
paper is to provide a few observations of my own
with respect to the linguistic potentials of nohuman
primates. The provision will be solely based on the
facts and knowledge of phonetics and gross anatomy
(1). In the course of this discussion, I will also
make use of the material presented in the preceding
papers in order to integrate the volume, although
the summary and evaluation has been undertaken in
the final remarks by Hewes.

To begin with, let me refer the reader back to
"Sign Language and Culture," where I briefly described the place and value of sign language and the
culture of the deaf vis-à-vis the problem of language

acquisition in man and ape. I should like to point
out that there are anatomical reasons (i.e., the
employment of the brachial apparatus, rather than
the vocal apparatus, the latter being quite different in man and ape [2]) to believe that to simply
speak, in the sense of "variability within fixed
frames" suggested by Voegelin, does not require
much brain capacity; something like an ape's brain
will suffice. However, there is a big difference
between Polly wants a cracker (or any other substitution needed) and Einstein invented the theory
of relativity (or any other substitution needed),
even though the two frames are essentially alike.
In this connection, I should say that the difference
lies not so much in the syntactic construction as
in the semantic comprehension. That is, it does
not take much intelligence to know what a cracker
(or nut or even banana) is, but a great deal of intelligence is required to understand the theory of
relativity. My contention, then, is that to talk
at the level of Polly wants a cracker (with its variability) or at the level of Einstein invented the
theory of relativity (with its variability) without
having the slightest idea of the theory of relativity (a situation that we will invariably en-
counter, if we learn a foreign language in pattern-
practice drill) does not require a brain of high
intelligence like ours; a brain of the calibre of
a chimpanzee or a gorilla should be good enough.

The fact that apes have not been able to talk,
then, is not because they do not have the kind of
intelligence needed (i.e., the right neurophysiological mechanisms) to talk, but because they do
not have the right vocal apparatus (3). An analogy
would be that a cellist should not be judged as not
being a virtuoso, if he is given a violin to play.
By analogy, given the right vocal apparatus apes
could talk, at least as much as they have been able
to sign using the brachial apparatus they have, and
 heir signing is much more than being able to vary
a frame like Polly wants a ... In other words, if
the apes' vocal apparatus were as good as their
brachial apparatus, they could speak just as well
as they can sign. On the other hand, if man's vocal apparatus were only as good as the apes', no
matter how great man's brain capacity (with its
superior size and association areas in the cerebral
cortex, as was suggested by Lancaster [1968:446-56],
albeit incorrectly), he would not be able to speak
beyond those few words that some apes have managed

to acquire. To follow up this line of argument,
then, let me first explain in some detail the anatomical differences of the vocal apparatus between
man and ape.

Five sagittal sections of head and neck of the
Primate Order are illustrated in the Figures (1-5).
A sixth figure for comparison shows only a part of
the oral and pharyngeal cavities and the larynx of
chimpanzee. The gorilla's sagittal section of head
and neck would look much like that of chimpanzee (4).
Note the marked differences in the oral cavity between man and the other nonhuman primates. Note also the difference in the size of the tongue and its
musculature (genioglossus, geniohyoideus, longitu-
dinalis linguae, and transversus and verticalis linguae) (5). Note further the detachment of the epiglottis from the velum in man versus the attachment
or overlaps in the others (6).

These differences suggest that while man has
plenty of room to maneuver in the oral and pharyngeal cavities, little room is available to the apes.
How much such differences in maneuverability mean
in terms of speech production is, of course, the
crucial question. The answer can best be shown in
the phonetic diagrams which follow.

Figure 7 indicates the various contours of
the tongue in relation to the pharynx and velum within the oral cavity. Notice the space needed for
the movement of the tongue in order to form the various vowels. Figure 8 indicates the raising of
the back of the tongue and the closure of the nasal
pathway by the velum to produce back consonants,
such as [k]. The raising and closure are synchro-
nized for oral consonants but may be separately articulated for the production of nasal consonants,
e.g., [ŋ]. I should also add that raising the back
of the tongue is an essential asset to the repertoire of sound production in human speech, a kinetic
movement which nonhuman primates cannot do because
of the structural limitations in their apparatus.
This is evidenced by the fact that the raising is
accomplished at the expense of the lowering (or separation) of the epiglottis from the velum.

To further support this argument, let me point
out that there are only two or three chimpanzees
that were trained to produce some speech sounds, including Sachiko in Japan. The most famous was Viki

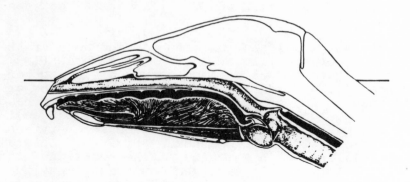

Figure 1.   Tupaia Minor. From E. Lloyd
DuBrul, Evolution of the Speech Apparatus,
1958.   Courtesy of Charles C Thomas, Publisher,
Springfield, Illinois.

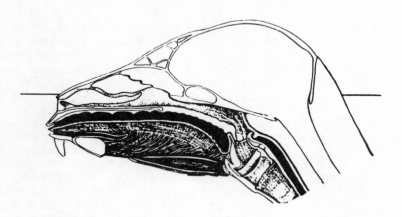

Figure 2.   Lemur Rufifrons.   From E. Lloyd
DuBrul, Evolution of the Speech Apparatus,
1958.   Courtesy of Charles C Thomas, Publisher,
Springfield, Illinois.

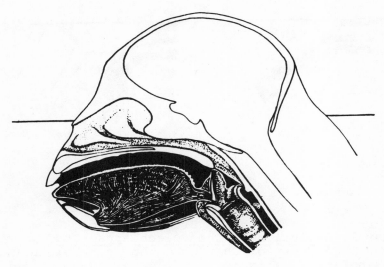

Figure 3.   Cercopithecus.   From E. Lloyd
DuBrul, Evolution of the Speech Apparatus,
1958.   Courtesy of Charles C Thomas, Publisher,
Springfield, Illinois.

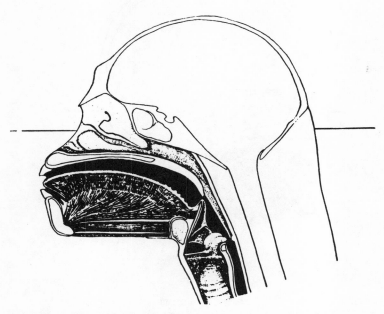

Figure 4.   Hylobates Lar.   From E. Lloyd
DuBrul, Evolution of the Speech Apparatus,1958.
Courtesy of Charles C Thomas, Publisher,
Springfield, Illinois.

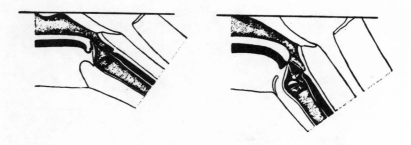

Baboon   Chimpanzee

Figure 5. Pan (right). From E. Lloyd
DuBrul, Evolution of the Speech Apparatus,
1958. Courtesy of Charles C Thomas, Publisher,
Springfield, Illinois.

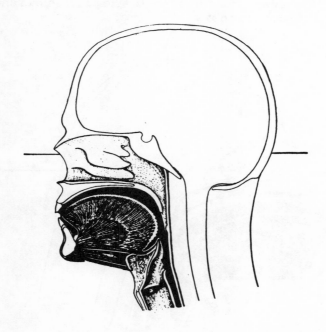

Figure 6. Homo Sapiens. From E. Lloyd
DuBrul, Evolution of the Speech Apparatus,
1958. Courtesy of Charles C Thomas, Publisher,
Springfield, Illinois.

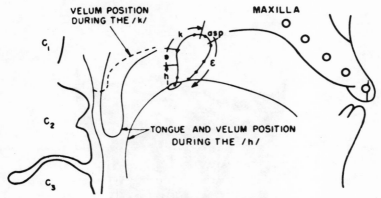

Figure 7. Tongue and Velum Positions in
the Productions of two Back Consonants. From
Joseph S. Perkell, Physiology of Speech Production: Results and Implications of a Quantitative Cineradiographic Study, 1969. Courtesy
of The M.I.T. Press, Cambridge, Massachusetts.

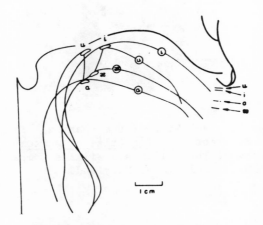

Figure 8.   Tongue Positions for Vowels. From
Joseph S. Perkell, Physiology of Speech Production: Results and Implications of a Quantitative Cineradiographic Study, 1969. Courtesy
of the M.I.T. Press, Cambridge, Massachusetts.

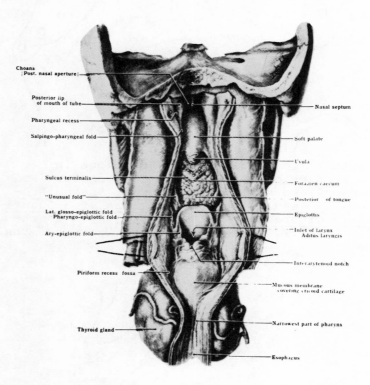

Figure 9. Back View of the Pharynx. From
J. C. B. Grant, Grant's Atlas of Anatomy, 1972.
Courtesy of The Williams & Wilkins Company,
Publishers, Baltimore, Maryland.

studied by the Hayes. Very few primatologists other
than the trainers heard <u>any</u> of those sounds produced
by at least one of those chimpanzees. If they did,
I doubt very much that they could describe in phonetic terms what the sounds were like; if they have
not, the chances are that they took the surface value of the orthography of the words reported by the
investigators. For instance, the Hayes reported
that Viki could pronounce the word "cup." By taking
the surface value of that word, without actually
hearing Viki's pronunciation, one is likely to be
misled, thinking that Viki produced a velar consonant represented by the English letter "c." But any
well-trained phonetician who has heard Viki's pronunciation will immediately say that what Viki produced is not a velar consonant, like [k], but something else. I possess two films on the training
programs of Washoe, Lana, and Viki. On examining
Viki's portion, I am convinced that what Viki produced is not a velar consonant for the word "cup"
but a glottal fricative, which is a friction noise
caused by the air passing through the vocal folds
when they are wide open, as if she were saying "hop."

Although they cannot raise the back of the
tongue, nonhuman primates can lock the nasal pathway with the epiglottis, so that there is a conti-
nous respiratory tube directly from the trachea into the nasal cavity, thereby by-passing the oral
cavity. The formation of such a tube is functionally important in two respects: (1) When food is
in the mouth, it can be masticated freely without
any danger of its going down the wrong pipe; (2)
prehension of the tongue is greatly enhanced, because the tongue is now freed from an additional
function which is to be imposed upon a human, as
will be explained below. Incidently, all mammals
except humans can form this direct respiratory tube.
And the increase in prehension as a derivative of
the function of this direct respiratory tube is more
striking among quadrupeds. For instance, a dog lap-
ping or a water buffalo eating grass against gra-
vity requires a prehensile tongue; but in the bi-
pedal human, prehension of this kind is absent.

Since the epiglottis in the case of man is no
longer high enough (Figure 9) to form a direct respiratory tube, a substitute must be found to do a
similar job, that is, the blocking of food from
going down the wrong pipe during the process of
mastication. (Note that the human respiratory tube

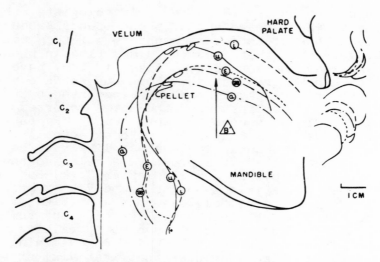

Figure 10.   Tongue and Lip Coordinations for the Production of Vowels.   From Joseph S. Perkell, <u>Physiology of Speech Production: Results and Implications of a Quantitative Cineradiographic Study</u>, 1969.   Courtesy of the MIT Press, Cambridge, Mass.

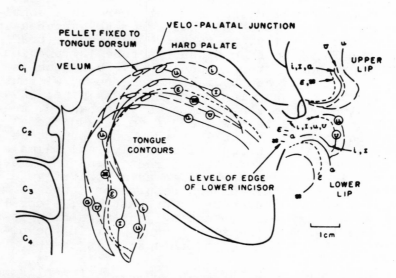

Figure 11.   Positions and Coordinations of Tongue and Lips for the Productions of Various Vowels.   From Joseph S. Perkell, <u>Physiology of Speech Production: Results and Implications of a Quantitative Cineradiographic Study</u>, 1969. Courtesy of the MIT Press, Cambridge, Mass.

now bifurcates, one going through the nasal cavity and the other the oral cavity.) This substitute is the back of the tongue; it does a reasonably good but occasionally faulty job. But this function is added at the cost of losing a considerable amount of prehension, because the musculature has had to change in order to accommodate the added function of the tongue. Of course, swallowing in primates also involves the tongue. But the point here is that other than swallowing the human tongue can in addition raise its back to block food from going down the wrong pipe, whereas the chimpanzee's tongue cannot do so (because there is no need for this function).

Now, to understand the function, try the following tricks. When you have food in the mouth to chew, breathe through the nostrils at the same time. You will soon discover that so long as the back of the tongue is raised against the palate, while you chew food, breathing is smooth. But the moment the closure is broken, as in the case of talking with food in the mouth, you immediately risk the danger of food particles going into the larynx. If you are not convinced with food in the mouth, try water.

What I have just said comes down to the fact that man has gone one step further to make good use of the added function of the tongue, i.e., the raising of the back of the tongue, for sound production purposes; in the case of nonhuman primates, since the anatomical structure has remained intact, the raising of the back of the tongue cannot be accomplished or if it is done accidently, it is nonfunctional (7). Hence, a wide variety of sounds involving this mechanism cannot be produced by nonhuman primates.

Figures 10 and 11 illustrate the various contours of the tongue in relation to the lips and the position of the lower incisors for the vowels. Note that the lips must be coordinated closely with the tongue to produce the vowels needed. Since the tongue of nonhuman primates is highly restricted in elevation and depression, as was already illustrated above, Voegelin was wrong in saying that "the lips of chimpanzees are more extensible than those of man, and should therefore function exceptionally well in producing rounded vowels." By the same token, Lancaster was in error, when she wrote:

"It will be noticed that no mention has been made
of the anatomical mechanisms used in making sounds.
This deliberate omission comes from the conviction that
no evolution in the mouth, tongue, or larynx was necessary to initiate the origin of human language"
(1968:453 fn. 5).

In this connection, I should add that while it
may seem quite difficult to see the important relation between the evolution in the mouth, tongue,
and larynx and the origin of human (oral) language,
it should not be too difficult to realize that the
change of modality (from the auditory mode to the
visual mode), coupled with the change of apparatus
(from vocal to brachial), has enabled not only chimpanzees but gorillas to acquire ASL, a human language, at a rate that is astonishingly comparable
to that of human children learning the same language. This fact, which is the difference between
all the words (three or four) that Viki was able to
produce after a couple of years (three years according to my films) of hard training and the signs
(close to three hundred) that Washoe or Koko has
mastered (not merely as isolated lexical forms but
in actual combinations as utterances) over the same
length of time, can hardly be explained, if only
the brain capacity (and nothing else) matters in
the evolution of human language.

In the same vein, I should also point out that
upon reexamining the films mentioned earlier, I
have discovered that those few words the chimpanzee
Viki produced have one thing in common; they lack
voicing. (This fact was "verified," albeit unknow-
ingly, by Tsuneya Okano, a psychologist at the University of Shizuoka, who spent six months to teach
Sachiko to say "mama," using all the means at his
disposal. He gave up the project, because Sachiko
whispered [mAmA] instead.) What this means is that
Viki's "papa" and "mama" (or Sachiko's "mama" for
that matter) sounded like [pApA] and [mAmA], whereby each [A] after [p] or [m] is just an opening of
the mouth with a little air friction in contrast to
an ordinary pronunciation by a human which should
be a voiced vowel. In other words, what Viki or
Sachiko produced in [A] above is a voiceless "vowel,"
a fact suggesting that neither animal learned to
vibrate the vocal folds in coordination with the
articulation of the consonants. All this could be
attested by using an x-ray motion-picture camera,

as was done in the case of a human subject. Unfortunately, no investigator of chimpanzee language behavior has done this, even though the x-ray technique in phonetics has been in existence for quite some time. I do not say that Viki or Sachiko could not vibrate their vocal folds, an act that would produce voicing. What I do say is that voicing and the articulation of other sounds must be well coordinated, a behavior that cannot be observed from the outside when learning a voiced sound. Perhaps Viki and Sachiko did not learn voicing because they could not observe it visually.

Now, without utilizing phonetic principles, without applying these to Viki's pronunciations (or Sachiko's), and without hearing her pronunciation of the word(s) she had learned, it would be very difficult to argue against my statement that nonhuman primates are not able to raise the back of their tongues. To convince me, to the contrary, one needs to transcribe Viki's pronunciations in phonetics or put them in a machine, like the sound spectrograph, to provide evidence of a velar consonant.

Similarly, Figure 12 illustrates that a variety of tongue contours must be in close coordination with the lips to produce other consonants than the back series. But again, nonhuman primates simply do not have enough room in the oral and pharyngeal cavities for maneuvering the tongue in order to produce those sounds, a disadvantage (or is it a disadvantage?) that is reinforced by the limitations in musculature of the tongue.

The aforementioned differences in vocal apparatus between man and ape are supported by van Lawick-Goodall (1968:318) in her description of the chimapnzee's common calls and facial expressions and by Lieberman's experimental acoustic analysis (1972). Note also that of all the sounds observed and described by van Lawick-Goodall, not a single one could be attributed to the raising of the back of the tongue or the coordination of the lips and the tongue.

Inability to produce human-like sounds, because of anatomical limitations, should by no means automatically give rise to the hitherto common but grossly mistaken assumption that apes do not have the neurological mechanisms and cognitive capability

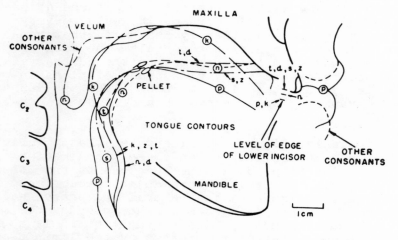

Figure 12. Positions and Coordinations of Tongue and Lips
for the Productions of Various Consonants. From Joseph S.
Perkell, <u>Physiology of Speech Production: Results and Implications of a Quantitative Cineradiographic Study</u>, 1969.
Courtesy of the M.I.T. Press, Cambridge, Massachusetts.

for language, as was suggested by Bender (and perhaps implied by Lancaster as well). For one thing, the neurological mechanisms and cognitive capability of lower primates need not be measured on the basis of oral language, any more than a cellist's musicianship could be properly judged in terms of his playing the violin. For another thing, the failure to teach chimpanzees in the past to speak a human oral language is due for the most part to the fact that the sounds of such a language (e.g., English or Japanese) that were imposed upon them have been derived from the anatomical mechanisms of man's vocal apparatus; never once have apes been taught to speak an oral langauge whose sounds were derived from the vocal apparatus of nonhuman primates (e.g., using a computer with an artificial vocal apparatus that could produce the sounds of nonhuman primates). Should there be such a simulated language, which could very well be simpler than any human oral language, I believe nonhuman primates, chimpanzees in particular, could learn to speak it.

Fortunately, however, there is convincing evidence, as presented in the preceding chapters, to support the fact that at least chimpanzees and gorillas do have neurological mechanisms and cognitive capability for language, if measured in a different mode, using sign language. And if Lancaster (1968: 453) is correct in saying that the ability to name objects provided man with a way to refer to his environment and to communicate information about his environment to other men, then Washoe (who according to Fouts — personal conversation on March 23, 1977 — referred to her fellow chimpanzees as "insects" and "savages") and Koko and many other nonhuman primates that have acquired ASL certainly have the ability to name objects with ASL to refer to their environments and to communicate information about their environments to humans.

There is one important question that remains unanswered, however. That is, have chimpanzees or gorillas, in their natural habitat, ever developed their own language, be it oral or manual or a combination of both, which could be regarded as a part of their own culture? Or, simply, do the great apes have a culture of their own? Or, in the light of Frishberg and Caccamise et al., do the great apes invent their own code for communication in their own culture? According to Lancaster, the answer seems negative:

"The ape makes many sounds. It sees much as we do.
Its facial expressions are rich in number and meaning.
But the structural basis for combining these abilities
into meaningful language is lacking" (1968:453 fn. 5).

"The more that is known about the communication
systems of nonhuman primates the more obvious it is
that these systems have little relationship with human
language, but much with the ways human beings express
emotion through gesture, facial expression, and tone
of voice ... It is human language, a highly specialized
aspect of the human system of communication, that has
no obvious counterpart in the communication systems
of man's closest relatives, the Old World monkeys and
apes" (1968:446).

But it seems very strange that while many primatologists, including Lancaster, agree that the
communication system of nonhuman primates is complex (probably implying that they do not really
know what to make of it) they assert also that
there is no grammar in such a system, or something
to that effect, as if they had already thoroughly
understood the system. Here are some more passages
from Lancaster, paraphrasing Marler (1965) and
Altmann (1965) or quoting Marler (1965) directly,
that pertain to my suspicion:

"One of the most significant generalizations, which
he (Marler) found demonstrated over and over again, is
that the communication systems of higher primates are
extraordinarily complex ... and that they rely heavily
on multimodal signals. A vocalization, a gesture, or
a facial expression in itself usually does not represent a complete signal, but is only a part of a
complex constellation of sound, posture, movement, and
facial expressions. Parts of such a complex pattern
may vary independently, and may help to express changes
in intensity or level of motivation. Sometimes olfactory elements are also present in the signal pattern but in monkeys and apes and in man the senses of
vision, audition, and touch are important in receiving
communicative signals" (Lancaster 1968:441-2).

"Another important generalization that has emerged
from the field studies of higher primates is the major
role that context plays in the total meaning of the
signal pattern. The receiver of a signal is presented
with an extremely complex pattern of stimuli. Not only
are the posture, gestures, vocalization, and facial
expression of the signaling animal important, but also

the total context of that pattern is an essential part
of the message. The immediately preceding events, the
social context, and the environmental context, all play
major roles in the way a signal is received, interpreted, and responded to. A threat display given by
a juvenile may be ignored in one context, whereas if
the same display is given again when he is near his
mother, and if she shows some interest in what he is
doing, it may produce an entirely different response
in the animal receiving the threat. The major function
of context in the total meaning of the signal makes
the study of primate communication systems very difficult. Responses to a signal pattern may seem highly
variable and erratic until a large number has been
sampled and the relevant aspects of the varying contexts of the signal have been taken into account"
(Lancaster 1968:442-3).

"Marler (1965:584) in his review of research on
primate communication systems concluded: 'Environmental
information, present or past, figures very little in
the communication systems of these animals, and a major
revolution in information content is still required
before the development of a variety of signals signi-
fying certain objects in the environment and a system
of grammar to discourse about them can be visualized'"
(Lancaster 1968:444-5).

Since I am not a primatologist, I have no way
to offer a counter example based on first-hand data.
As a linguist, however, I am under the impression
that what is meant by "structural basis," "meaningful language," and "a system of grammar to discourse"
is reminiscent of something high school teachers of
Latin used to say and may still say about English
and Chinese; that is, "English and Chinese do not
have a grammar." What they mean by this is that
there is no conjugation for verbs and there is no
declension for nouns and so on. In other words,
they try to look for patterns that resemble the
grammar of Latin in English and Chinese, and when
they cannot find such patterns, they conclude,
"English and Chinese do not have a grammar."

Whether or not primatologists have made a similar mistake in their generalizations is anybody's
guess. But the danger of making the same mistake
exists, judging from what has been said so far.
Therefore, care must be taken to avoid it. For instance, instead of looking for a system that resembles human oral language, one may attempt to

examine the communication system of chimpanzees in
its own right, whereby the posture, gestures, vocalization, and facial expression of the signaling
animal plus the total context of that pattern may
very well constitute not just a complex constellation of sound, posture, movement, and facial expressions but a rather organized whole with meaningful, internal, structural relationships among
the various components.

In sign language, interestingly enough, facial
expressions are part and parcel of the grammar which
is also highly context-sensitive. The same is true
of oral language. More and more linguists are becoming aware that, contrary to earlier assumptions,
language is not a self-contained entity: the context plays a highly significant role in the communication system of humans, much more so than we
have thus far realized. The posture, gestures, vocalization, and facial expression of the signaling
human individual must be congruent in order to convey the appropriate meaning; and the immediately
preceding events, the social context, and the environmental context all play major roles in the way
a signal is received, interpreted, and responded to.
Does this sound familiar? Perhaps it is the complex constellation of the various components (sound,
posture, movement, gesture, and facial expression)
in close relationship with the actual context of
situation that should be regarded as "the origin of
language." If so, this constellation gradually developed and differentiated into its present form in
humans; the vocal component and the manual component
have become more specialized, although still in
close relationship with the situational context,
while accompanied by concomitant gesticulation,
kinesics, facial expressions, and other body movement in the same consellation. Thus, if one is to
look for the original form of "language," he will
not find it in the vocal component employed by nonhuman primates (nor in the fossil remains of the
Neanderthal, for that matter). However, if he
tracks down the origin of the "consellation," he
will find abundant traits and traces in the communication systems of living nonhuman primates.

One actual example will suffice to illustrate
this point. Many years ago, I attended an International Conference held in Boston. A friend invited me to a restaurant for dinner. When we arrived, the parking lot was full. Seeing two

entrances, we took the one closer to us. But as
our car entered the parking lot, we saw another
car coming from the opposite entrance, both drivers
now looking for a place to park. Just then, a car
pulled out of its parking space, giving hope to
both waiting drivers. However, the third car backed
up in such a way that it blocked the pathway of the
other car. Our car had an advantage, and my friend
quickly occupied the parking space. As we got out
of our car, the driver of the other car, a woman,
also got out of her car, walked towards us, and as
soon as she was near enough, stopped and said,
"Thanks a lot!" The tone of voice could in no way
be mistaken for expressing gratitude. Then she
walked away.

Now, imagine her posture and gesture, while
talking to us, and her facial expressions (angry or
happy?) Being a woman, she might have expected
"Ladies first," which we ignored, a behavior that
was not congruent with the social norm ten years
or so ago. Did her vocalization, gesture, posture,
or facial expression in itself represent signals
to us? Each separate action was a part of a complex constellation of sound, posture, movement,
facial expressions, and gesture, from which we got
the message.

If this example is illustrative of the communicative constellation in American society, do we
have a right to say that chimpanzees or gorillas
do not have a culture of their own? They show similar behaviors within their own social group (e.g.,
chimpanzees in van Lawick-Goodall 1968 and gorillas
in Fossey 1970 and 1971 and Schaller 1963), even
though we cannot make out what they are "saying" to
each other in the total constellation. In any
event, the questions raised above await answers
that must come from careful field observations (8).

Notes

1. While I am a trained phonetician, I also
have studied anatomical sciences at SUNY (Buffalo,
N.Y.).

2. The anatomical reasons refer to the structural differences in the vocal apparatus between
man and ape.

<u>3</u>. Intelligence is traditionally defined as "the ability to learn."

<u>4</u>. For sagittal section of head and neck of the gorilla, see Negus (1949).

<u>5</u>. Two important muscles are missing from the sagittal sections, namely, styloglossus and hyoglossus, because they are not in the midline. For the tongue to function properly in sound production, moreover, other muscles in the neck region are needed, such as stylohyoidues, digastrics, and mylohyoideus, as well as muscles of the infrahyoid, muscles of the palate, and muscles of the pharynx.

<u>6</u>. The position of the epiglottis in a neonate relative to the uvula is pretty much the same as in a chimpanzee. The epiglottis, in the case of a human baby, begins to descend at the age of six months; it reaches the "adult" position when the baby is about two years old. Without this descending, it would be very difficult for an infant to produce one-word not to mention two-word utterances at the age of two.

<u>7</u>. By "accidentally" is meant that under some extreme conditions the back of the tongue may hit the velume without the intention of producing any sound. This situation, by analogy, may be likened to a person jumping from a building that is three- or four-stories high, when there is a fire in the building.

<u>8</u>. I am grateful to Gordon W. Hewes of the University of Colorado for material improvement in this chapter.

References Cited

Altmann, S. A.
    1965   "Sociobiology of Rhesus Monkeys, II, Stochastics of Social Communication" <u>Journal of Theoretical Biology</u> 8.490-522.
    1967   "The Structure of Primate Social Communication" in S. A. Altmann (ed.), <u>Social Communication among Primates</u>, pp. <u>325-62</u>, Chicago: The University of Chicago Press.

Bender, M. Lionel
    1976 "In Defense of Linguistic Innateness: Re-
          ply to Peng" Language Sciences, August.
          pp. 19-20.

Fossey, Dian
    1970 "Making Friends with Mountain Gorillas"
          National Geographic 137(1).48-67.
    1971 "More Years with Mountain Gorillas"
          National Geographic 140(4).574-85.

Gardner, B. T. and R. A. Gardner
    1971 "Two-way Communication with an Infant
          Chimpanzee" in A. M. Schrier and F.
          Stollnitz (eds.), Behavior of Nonhuman
          Primates, 4, pp. 117-83, New York:
          Academic Press.

Lancaster, Jane B.
    1968 "Primate Communication Systems and the
          Emergence of Human Language" in Phyllis
          C. Jay (ed.), Primates, pp. 439-57, New
          York: Holt, Rinehart, and Winston, Inc.

Lieberman, Philip
    1972 The Speech of Primates, The Hague: Mouton.

Marler, P.
    1965 "Communication in Monkeys and Apes" in
          I. DeVore (ed.), Primate Behavior: Field
          Studies of Monkeys and Apes, pp. 544-84,
          New York: Holt, Rinehart, and Winston,
          Inc.

Negus, V. E.
    1949 The Comparative Anatomy and Physiology
          of the Larynx, London: W. Heinemann Me-
          dical Books. (This is a condensed and
          rearranged version of the author's The
          Mechanism of the Larynx published in
          1929.)

Schaller, George B.
    1963 The Mountain Gorilla, Chicago: The Uni-
          versity of Chicago Press.
    1964 The Year of the Gorilla, Chicago: The
          University of Chicago Press.

van Lawick-Goodall, Jane
    1968 "A Preliminary Report on Expressive Move-
          ments and Communication in the Gombe

Stream Chimpanzees" in Phyllis C. Jay (ed.), Primates, pp. 313-74, New York: Holt, Rinehart, and Winston, Inc. Also in Phyllis Dolhinow (ed.), Primate Patterns, 1972, pp. 25-84, Holt, Rinehart, and Winston, Inc.

1971   In the Shadow of Man, Boston: Houghton Mifflin Company.

Comments and Remarks

Gordon W. Hewes

A Common theme pervades these seven papers;
namely, the notion of speech as the only true form
of language has obscured some important linguistic
understandings. Although writing is a form of language in the visual mode, linguists have generally
insisted that writing is little more than a secondary, derivative version of spoken language, not
worthy of much theoretical attention. The sign
languages of the deaf and of certain hearing communities have been even more thoroughly disparaged as
crude makeshifts or jargons, even less deserving of
serious analysis. The participants of this Symposium have been working with languages in the visual
mode, and several of them have had as their principal subjects not human beings but anthropoid apes.
We are, thus, dealing with topics which the majority of linguists have held to be marginal, irrele-
vant, or even frivolous. Professor Peng's paper
touches on many of these points, and bridges the
gulf which is supposed to separate language capable,
speaking mankind and mute, languageless apes.

Peng, discussing the problem of the relationship of language to culture, mentions Edward B.
Tylor (1832-1917), whose famous definition of culture is often the starting point of any argument
over what is or is not to be included in that concept. Tylor was deeply concerned with language in
relation to culture, or to the "rest of culture,"
but, more to the topic of our Symposium, devoted
much attention to manual sign languages both of the
deaf and of various tribal peoples; moreover, he
seriously considered the possibility that human language might have originated in gesture rather than

on the vocal channel. Garrick Mallery (1831-1894),
who published the first major work on the sign language of the North American Plains Indians in 1881,
in the first volume of the annual reports of the
Bureau of American Ethnology, was also an influence
on the thinking of late 19th and early 20th century
anthropologists, thanks to the prestige which the
Bureau then enjoyed in anthropological circles.
Despite these auspicious beginnings, anthropologists
and anthropological linguists, in the best position
to observe and record sign languages in ethnographic contexts, came to regard the entire topic as a
trivial ethnographic curiosity, without important
theoretical implications. As the field of linguistics gained in apparent rigor and theoretical so-
phistication, cultural anthropologists came to look
upon it as an ideal model toward which the whole
science of culture should turn for guidance. The
once impressive ediface of modern linguistics has,
however, been collapsing under the assaults of a
paradigm shift, and the behaviorist psychological
model on which earlier 20th century linguistics had
been patterened has likewise been very strongly
challenged. Both the current revival of interest
in sign language systems, and the unexpected emergence over the last decade of studies of the language capabilities of apes are part of the demoli-
tion process underway against the former paradigms.

To Peng's point that some scholars have
claimed that if two or more subject-areas are part
of some larger field, they must, therefore, be
amenable to the same analytical methods, I must say
that this view has not impressed me. Many anthropologists, overawed by the accomplishments of so-
called structural linguistics, and at the same time
comfortable with all of the implications of the assumption that language is a "part" of culture, have
indeed been led to force the analysis of ethnographic data into a linguistic methodology. At the
same time, while most anthropologists would agree
that music, animal husbandry, and higher mathematics are "parts" of some or many cultural systems,
they would probably acknowledge that the methods
and theories of cultural anthropology do not begin
to suffice for the analysis of these three systems.

Peng notes that sign language users tend to be
a low status, particularly if they are deaf. Profound deafness is a serious disability, and one

which strongly affects communication, education,
etc., and it is probably not surprising that using
a manual sign langauge is regarded as a symptom of
ineptness or even mental backwardness, quite un-
justifiably. When to this are added the inade-
quacies, in most countries, of education provisions
for the deaf, and the determined effort by most of
their professional teachers for the past century
and more to totally depreciate gestural communication in favor of the often much less effective "oral"
method, as well as the restricted occupational
choices offered to the deaf, it is no wonder that
either the use of a sign language or the serious
study of it is not highly regarded. When to these
considerations one adds the note that various primitive peoples employ sign languages, the "primitive" status of all such languages is made very
clear, and those who deal with them are put on the
defensive.

     We should expect to find universals of various
sorts in sign language systems. Dr. Peng provides
some interesting examples of parallels between ASL
and JSL (Japanese Sign Language). Nevertheless,
some linguists appear to be disturbed by the possibility of significant sign language universals, as
if the next step were the inevitable revival of the
"psychic unity of mankind" notions of the 19th century, some of which were neither so naive nor so
stupid as the early 20th century anthropological
reaction made out. There are many ironies here-
abouts. The same linguist who may bridle at the
suggestion of the existence of sign langauge universals may accept without demur the dogma that all
natural spoken languages are absolutely equivalent
with respect to the ease of learning them. Other
than the fact that most normal children manage some-
how to master their native tongues in a decade or
so, the hypothesis of equal learnability has never
been systematically tested. Yet, if I were to venture to suggest that manual sign languages might be
easier to acquire than spoken languages, at some
acceptable criterion of fluency, this would be held
to violate the principle of the equivalent difficulty of all spoken languages. As it happens, a few
scattered studies seem to indicate that young children of sign langauge using parents begin to acquire
signing well before hearing children begin to exhibit significant control of speech, which certainly
casts some doubt on the "principle" just mentioned.

Peng's discussion also suggests that there may be some communicational advantages of manual sign language over speech, especially with respect to transmission of spatial and movement information. Beyond this, there is much evidence to show that the visual channel, at least in the higher primates, has a far higher informational capacity than the auditory channel, including a much larger absolute number of afferent nerve fibers. The advantages of the visual mode for information handling have been set forth by J. Bertin (1969) among others, in connection with some fundamental differences between linear and graphical presentations of data. While part of the superiority of a sign system in the visual mode lies in the possibilities for explicit iconicity (for the shapes of objects or their movement) three dimensional space, as usable in a gesture language, can also convey many kinds of iconized abstractions. The history of science and technology over the past few centuries has witnessed an explosive expansion of graphics to convey iconized abstract relationships. While most, if not all, known natural spoken languages employ body-and body-part metaphor, sign languages may have an advantage in the directness with which such metaphors can be represented. Against such views, there are those who believe that high iconicity interferes with communication or understanding. To be sure, prior to the invention of printing, and some centuries later, photography and many other ways of reducing the amount of labor required for iconic representation, the production of iconic messages (in painting, drawing, sculpture, and the like) was indeed very slow and, at its higher levels, demanding skills present in only a small proportion of any population.

Had Dr. Peng and I had more time for comments, we might have said something about the high frequency of representations of human hands performing various actions in the oldest forms of Chinese written characters, later glossed in the <u>Shuo Wen Chie Tzu</u>. Is it conceivable that in its earliest formative stages, Chinese writing reflected some of the informational advantages of a visual-gestural communication system?

In quoting Carl Voegelin, a distinguished linguist, to the effect that language-like behavior will never be developed in apes, Peng reminds us not only of the hazards of such dogmatic assertions,

but also of how prevalent scholarly biases might in-
hibit promising lines of research. Had the Gardners
and David Premack been trained in linguistics, rather than in psychology, it seems quite clear that
they would never have ventured into the field of
teaching apes to acquire language or "language-like
behavior." Finally, Peng has reviewed for us some
of the anatomical reasons why apes appear to be unable to talk, although as Fouts and later Patterson
will tell us these vocal tract features do not prevent the same apes from developing some <u>receptive</u>
capacity for speech.

Dr. Frishberg strongly supports the view that
sign languages are in most respects formally comparable to spoken ones, aside from certain differences imposed by the different physical channels
for exchanging visible signs and auditory signs.
She clears up several popular misconceptions about
sign languages, beginning with the widely held notion that there is one universal or international
sign language. However, the nonexistence of such
an international sign language does not mean that
its creation and adoption would be impossible in
principle. There are in fact international groups
of the deaf working on such a system. The reasons
which would lead it to fail lie in the sociolinguistic realm, just as they do for would-be international spoken and written auxilliary languages like Esperanto. The relative failure of Esperanto lies
not in the fact that its creator, Zamenhof, chose
to base it entirely on certain European languages,
but largely because there is likely to be a greater
personal pay-off, if one is willing as an adult to
study another language, in acquiring one of the major world languages, with all of their well-known
irregularities.

Frishberg deals with some possible sign language universals, such as the apparent tendency to
confine signing to a space in front of the body,
between the top of the head and the waist, and to
employ only a limited set of possible hand shapes,
placements, and kinds of movement, somewhat on the
analogy of phonemes in spoken language. Although
it is clear, from the available documentation, that
ASL and its late 18th century French ancestor, which
was the "reformed" sign language of the Abbé de
l'Epée (1712-1789) carried to New England by T. H.
Gallaudet, had considerably more mime and iconic
signs in the past than it has now, Frishberg may go

too far in her assumption that a fully mature sign
language would be practically without face-value
iconicity. Even spoken languages, it seems, contain
various kinds of iconicity, such as sound symbolism
(as well as certain kinds of syntactic iconicity),
and their development has probably been much more
continuous than that of any known sign language.
In her discussion of the wide variety of ways of
signing the first ten numbers, some basic iconicity
remains. No manual sign language is likely to find
it convenient to represent "three" by either raising or depressing just one of the ten fingers, on
the ground that three is the square root of nine,
or "zero" by five fingers because five from five
equals zero. Our intuitive rejection of these possibilities would appear to rest on what might be
called the psychic unity of mankind.

Concerning the two columns of non-alphanumeric
symbols which Frishberg discusses, the first (left-
hand) one of which has ASL "names" for each symbol,
and the second (right-hand) column of which has
none, I can only say that this does not surprise
me at all. The items in the first column are most-
ly common symbols used in English written compos-
ition, and are of the kind that would probably be
regularly referred to in the schoolroom. Those in
the second column are less common, and many of them
seem to lack names in spoken or written English.
Thus, although I have been literate for many years,
and I know its meaning(s), I have no name at all
for the sign #. Also, on my typewriter, I find the
symbol @; I know it can mean 'at', as in the expression "1 doz. @ $1.25" but I am not at all sure
that a professional printer would refer to it as an
"at" sign. I was shocked not long ago to find that
a college student not only did not know what to call
the symbol & (it is an "ampersand") but the student
did not even know what it meant.

I am not especially impressed by the possible
existence of duality of patterning in sign language,
since analogous principles can be found in a great
many systems for coping with very large sets of
items, such as books in a library catalogue, or au-
tomobile license plates. I am impressed, however,
with the phenomenon which Dr. Frishberg herself
exemplified at the Denver AAAS Symposium; namely,
the simultaneous delivery of her remarks in normal
English speech and ASL plus fingerspelling. As she
later pointed out, such performances distort ASL,

which in such circumstances tends to follow English word-order and so on, but it is nevertheless the case that for the nonfingerspelled portions of her discourse, she was producing word-signs simultaneously in two very different ways. Presumably, a highly proficient Chinese could deliver a talk while simultaneously writing it down in Chinese characters, but I suspect that the spoken message would be markedly slowed down, compared to normal delivery. Possibly, a trained shorthand clerk or stenotypist could speak and write a message at normal speaking rates, but there the analogy with ASL signing breaks down; the components of shorthand or stenotype are phonetic and have a linear correspondence with the syllables of the spoken message.

Caccamise, Blasdell, and Bradley, in their joint paper, address another problem — the development of acceptable new technical signs for ASL. This is very much like the problem of introducing or creating new scientific and technical words into spoken languages, currently important for many developing countries. Language planners have several possible solutions. ASL may simply take over a needed term by fingerspelling it, or perhaps reducing it to a handier abbreviation. This resembles direct word-borrowing by a spoken or phonetically written language, where the loanword may be modified only to conform to the sound system of the language which is doing the borrowing. Another process is to determine the underlying meaning of the foreign term and, then, to create a native equivalent for it out of existing native roots or morphemes. German seems to prefer this procedure, as has Chinese. Modern Japanese has used both methods, and in addition to creating large numbers of new kanji-compounds for technical and scientific concepts, e.g., /jinruigaku/ for 'Anthropology', i.e., "man + variety + science," it has simply followed the phonetic approximation route: /arubaito/ from German "Arbeit"; /apa:to/ from the first two syllables of English "apartment." ASL has several possibilities but seems to prefer utilizing existing standard features combined in novel ways, perhaps influenced by what Peng calls "sign symbolism." Unlike Chinese or German, it does not seem to use the method of analyzing the meaning components of a foreign word and, then, use its own primes to form a new ASL compound (1). The Germans, confronted with the Neo-Greek compound "geology," translated its components

into "earth + science" and produced /erdkunde/.
Perhaps, the genius of ASL does not lend itself to
such easy formation of new terms because, if at all
possible, it tends to prefer a single holistic sign
to a two- or three-sign compound. But why is this?
The present ASL way of handling the formation of
new terms seems to be more like George Eastman's
coinage of the catchy "Kodak" for his box camera
and roll-film, a coinage in which we may suspect
some sound symbolism, as in so many trade-names not
formed out of pre-existent roots.

With Lyn Miles' paper, we shift to the apes,
although she is still also concerned with language
acquisition rates and stages in human children,
with whom the apes may be usefully compared. Basically, her question is whether apes exhibit language-
learning behavior sufficiently similar to that of
human children acquiring their first language to
suggest very similar central (brain) processing
functions. Apart from the obvious inability of
apes to learn to speak, for reasons already explained by Peng, if language is presented to apes
and human beings on the visual-gestural channel,
no striking differences appear, at least as far as
any apes have gone in language mastery.

Although it seems very likely that the maximum
linguistic achievement of apes exists, we do not
now (as of 1977) know even the approximate upper
limits of their capacities in this realm of behavior.
Miles points to several problems which future research should try to solve. For example, is ASL
(or any of the other visual-motor languages being
inculcated in apes) being left-lateralized in the
brain, as it allegedly is in human beings? If not,
why not? Are there syntactical rules which most
human language-users eventually master (e.g., em-
bedments, as of subordinate clauses, use of the
passive voice, and the like) which will prove to be
beyond the linguistic skills of apes? Can apes
handle a highly inflected language system as well
as one based on strong word-order rules like English
or Chinese?

Roger Fouts attacks the persisting philosophical argument, which goes back practically un-
changed to Rene Descartes in the 17th century, that
man alone is capable of any sort of language, and
hence alone among terrestrial organisms possesses
conceptual thought. He also has been able to show

significant receptive acquisition of spoken English
in Pan, testable thanks to the fact that his chimpanzee subject possesses some facility in ASL. These
findings, coupled with more anecdotal reports of
others working with apes (among them, Francine
Patterson), challenge some of the notions of psycholinguistics with respect to the supposedly uniquely
human speech-sound decoding capacity, already somewhat clouded by the finding that chinchillas can
make some categorical distinctions in speech sounds
or their electronically produced equivalents.

Rumbaugh, Savage, and Gill describe their on-
going work with the chimpanzee Lana, and work now
starting with several other chimpanzees, including
members of the rare pygmy species, Pan paniscus.
The principal point of their joint presentation is
that language handling capabilities can best be regarded as part of a more general cognitive capacity,
most higly developed in Homo sapiens but also well
developed in the Pongidae, as the outcome of similar
evolutionary selective pressures. This is quite different from the notion, elaborated in the well-known
book of the late Eric Lenneberg, and incorporating
certain view of Noam Chomsky, to the effect that
language behavior is only possible in organism which
come equipped with cortically based language-process-
ing "devices," arising in the course of biological
evolution, but essentially without explanation.
Since several apes of different species seems to be
exhibiting at least a low level of language acquisition, the Lenneberg and Chomsky speculation would
require that they too were genetically equipped with
language acquisition "devices." In view of the lack
of anything closely resembling language among these
animals in their natural habitat, this would constitute a "pre-adaptation" of a surprisingly elaborate kind.

While apes have never been observed to employ
anything properly definable as a propositional language in the wild state, pygmy chimpanzees nevertheless appear to have the rudiments of a manual
gesture system, consisting of several different
hand and arm movements, apparently chiefly used to
direct the positioning or movement of others of
their species. In captivity, this capacity seems
to be mainly employed to direct or correct the
copulatory position of another individual. Limited
though this is, it is just the kind of basic deictic
gesture system that could have, over a great deal of

time, given rise to a rudimentary manual sign language, as I have suggested in some previous papers,
following a kind of glottogonic scenario developed
by the Vietnamese philosopher, Tran Duc Thao.   I
am, therefore, very pleased that Sue Savage is now
working with pygmy chimpanzees in a language experiment.

Francine Patterson's work with the young fe-
mal gorilla Koko, now supplemented by work with a
still younger male gorilla Michael, has dramatically
enlarged the scope of research into ape language
capacities. Despite the molecular biological data
which put Pan troglodytes closer to Homo sapiens
than any other higher primate, the work with Koko
makes it quite unlikely that this close biochemical
affinity renders chimpanzees more linguistically
capable than gorillas. Koko's lexicon had reached
244 different ASL signs, determined to be present
by a rigorous criterion of usage, and has so far
shown no indication of tapering off. The superior
performance to date of Koko could be due to her
probably somewhat larger brain (compared to the
average chimpanzee of her age), but is more likely
to be the result of very good personal rapport between Koko and her investigator. Further, Patterson
has had the advantage of being able to benefit by
the accumulated experience of all of the ape language investigators, and her work with Koko began
in Koko's infancy. Some impressive behavior not
strictly linguistic was reported for Koko, having
to do with remorse over misbehavior on the previous
day, and her ability to comment calmly on an event
which, when she experienced it, was emotionally
upsetting.

It seems legitimate to pool, as it were, the
results of all of the different ape language studies
so far, with the conclusion that what might appear
to be significant deficiencies in performance in
one animal or one line of experiment are offset by
accomplishments of subjects in other studies. Peng's
account of the linguistic potentials of nonhuman
primates seems supportive of this conclusion. Thus,
unless it should turn out that Lana's high level of
syntactic behavior would preclude her attainment of
a 244-item sign vocabulary, such as Koko has demonstrated, or vice versa, it is probably safe to assume that given suitable environments for language
acquisition, an ape, whether gorilla or chimpanzee
(and possibly also an orang) could achieve the

grammatical experties of a Lana, the lexicon of a
Koko, the spoken language receptivity of an Ally,
and so on.

Needless to say, I should like to see much more
along all of the lines presented by this Symposium,
and much more besides. We desperately need solid
descriptive information about a great many more
natural sign language systems around the world. And
this appears to be the spirit concerted by the authors of the sign language papers in this volume.
The systems will probably turn out to be built
around individuals or a few deaf people, in isolated
regions, like the sign language of Kagobai on Rennell
Island; in more developed countries, deaf children
are more likely to be placed in school situations,
where for some time to come their education will be
in the hands of teachers trained to suppress signing in the interest of oral language skills. It
would be interesting to know if sign languages developed for deaf individuals or families outlive
them, and enter into the "nonverbal" gestural re-
pertoires of speaking people and, of course, to
what extent the elements of such gestural para-
languages, prevalent among hearing people in a given
culture, are utilized by the deaf, and their relatives and friends.

With regard to ape language researches, a great
many exciting possibilities come to mind, almost
certainly far exceeding the practicalities of fund-
ing. Unfortunately, all of the ape language projects are very expensive. I have already wondered
whether ape-language learners would do as well with
languages based on the structure of Navajo or Eski-
mo as they seem to be doing with languages born in
a mainly English (or 18th century French) milieu,
though transformed into the visual mode. Some very
important experimental programs in ape psycholinguistics, neurolinguistics, and cognitive studies
in which language plays a critical role must be
undertaken.

Notes

1. In Peng and Clouse (1977), loan translation
which is supposedly absent from ASL is reported to
be a common process of borrowing in JSL with regard
to place name signs — Editor.

References Cited

Bertin, Jacques
    1969 "Graphique et Mathematique: Généralis-
          ation du Traitement Graphique de l'Infor-
          mation" Annales - Économies - Sociétés,
          Civilisations, 24 année, 1.70-101.

Peng, Fred C. C. and Debbie Clouse
    1977 "Place Names in Japanese Sign Language"
          in The Third LACUS FORUM 1976, Columbia,
          S. C.: Hornbeam Press, Inc.